W9-CMK-203

Also by Antonio Damasio

The Strange Order of Things

Self Comes to Mind

Looking for Spinoza

The Feeling of What Happens

Descartes' Error

Feeling

&

Knowing

Feeling
&
Knowing

MAKING MINDS
CONSCIOUS

Antonio Damasio

Pantheon Books, New York

Copyright © 2021 by Antonio Damasio
Illustrations copyright © 2021 by Hanna Damasio

All rights reserved. Published in the United States by
Pantheon Books, a division of Penguin Random House LLC,
New York, and distributed in Canada by
Penguin Random House Canada Limited, Toronto.

Pantheon Books and colophon are registered trademarks of
Penguin Random House LLC.

Grateful acknowledgment is made to Harvard University Press for
permission to reprint an excerpt of "The brain is wider than the sky"
from *The Poems of Emily Dickinson: Variorum Edition,* edited by
Ralph W. Franklin, Cambridge, Mass.: The Belknap Press of
Harvard University Press. Copyright © 1998 by the President and
Fellows of Harvard College, Copyright © 1951, 1955 by the President
and Fellows of Harvard College, copyright renewed 1979, 1983 by the
President and Fellows of Harvard College. Copyright © 1914,
1918, 1919, 1924, 1929, 1930, 1932, 1935, 1937, 1942 by
Martha Dickinson Bianchi. Copyright © 1952, 1957, 1958, 1963, 1965
by Mary L. Hampson.

Library of Congress Cataloging-in-Publication Data
Names: Damasio, Antonio R., author. Damasio, Hanna, illustrator.
Title: Feeling & knowing : making minds conscious /
Antonio Damasio ; illustrations by Hanna Damasio.
Description: First edition. New York : Pantheon Books, 2021.
Includes bibliographical references.
Identifiers: LCCN 2020034554 (print). LCCN 2020034555 (ebook).
ISBN 9781524747558 (hardcover). ISBN 9781524747565 (ebook)
Subjects: LCSH: Consciousness.
Classification: LCC BF311 .D3233 2021 (print) |
LCC BF311 (ebook) | DDC 153—dc23
LC record available at lccn.loc.gov/2020034554
LC ebook record available at lccn.loc.gov/2020034555

www.pantheonbooks.com

Jacket image: Rock Island Split 2019 by Wayne Thiebaud. © 2020
Wayne Thiebaud/Licensed by VAGA at Artists Rights Society (ARS),
N.Y. Photo courtesy Acquavella Galleries.

Jacket design by Kelly Blair

Printed in the United States of America
First Edition

1 2 3 4 5 6 7 8 9

For Hanna

"The life of a play begins and ends in the moment of performance."

—Peter Brook

Contents

The Substrate Counts · Loss of Consciousness ·
The Cerebral Cortices and the Brain Stem in the
Making of Consciousness · Feeling Machines and
Conscious Machines

Before We Begin

1

The book you are about to read has some curious origins. It owes a lot to a privilege I have long enjoyed and a frustration I have often felt. The privilege consists in having had the luxury of space when I needed to explain complicated scientific ideas using the large number of pages of a standard nonfiction book. The frustration came from talking to many of my readers, over the years, and learning that some ideas that I wrote about with enthusiasm—and that I had been keenest to have readers discover and enjoy—were lost in the middle of long discussions and hardly noticed, let alone enjoyed. My private response, on such occasions, has been a firm but always postponed decision: to write *only* about the ideas I most care for and leave behind the connective tissue and the scaffolding meant to frame them. In brief, do what good poets and sculptors do so well: chip away at the nonessential and then chip some more; practice the art of haiku.

When Dan Frank, my editor at Pantheon, told me that I should write a focused and very brief book on consciousness, he could not have anticipated a more receptive and enthusiastic author. The book you have in your hands is not exactly what he ordered, because it is not *only* about consciousness, but it comes close. What I could not have anticipated is that the effort of reconsidering and paring down so much material, would help me confront facts that I had overlooked and develop new insights about not just consciousness but related processes. The road to discovery is twisted, to say the least.

It is not possible to make sense of what consciousness is and of how it developed without first addressing a number of important questions in the universe of biology, psychology, and neuroscience.

The first of those questions concerns *intelligences and minds*. We know that the most numerous living organisms on earth are unicellular, such as bacteria. Are they intelligent? Indeed they are, remarkably so. Do they have minds? No, they do not, I believe, and neither do they have consciousness. They are autonomous creatures; they clearly have a form of "cognition" relative to their environment, and yet, instead of depending on minds and consciousness, they rely on *non-explicit competences*—based on molecular and sub-molecular processes—that govern their lives efficiently according to the dictates of homeostasis.

And what about humans? Do we have minds and only minds? The simple answer is no. We certainly have minds, populated by patterned sensory representations called images, *and* we also have the non-explicit competences that serve simpler organisms so well. We are governed by two types of intelligence, relying on two kinds of cognition. The first is the one humans have long studied and cherished. It is based on reasoning and creativity and depends on the manipulation of explicit patterns of information known as images. The second type is the non-explicit competence found in bacteria, the one variety of intelligence on which most lives on earth have depended and continue to depend. It remains hidden to mental inspection.

The second question we need to address deals with the ability to feel. *How are we able to feel pleasure and pain, well-being and sickness, happiness and sadness?* The traditional answer is well known: the brain allows us to feel, and all we need is to investigate the specific mechanisms behind specific feelings. My aim, however, is not to elucidate the chemical or neural correlates of one particular feeling or another, an important issue that neurobiology has been attempting to address with some success. My aim is different. I wish to know about the functional mechanisms that allow us to *experience in mind* a process that clearly takes place in the *physical realm of the body*. That intriguing pirouette—from physical body to mental experience—is conventionally attributed to the good offices of the brain, specifically to the activity of physical and chemical devices called neurons. Although it is apparent that the nervous system is required to accomplish that remarkable transition, *there is no evidence that it does so alone*. Moreover, the intriguing pirouette that allows the physical body to harbor mental experiences is regarded by many as impossible to explain.

In an attempt to answer the critical question, I focus on two observations. One of them relates to the unique anatomical and functional features of the interoceptive nervous system—the system responsible for signaling from the body to the brain. These features are radically different from those that can be found in other sensory channels, and although some of them have previously been documented, their significance has been overlooked. And yet they help explain the peculiar melding of "body signals" and "neural signals" that decisively contributes to experiencing the flesh.

Another pertinent observation concerns the equally unique relationship between the body and the nervous system, specifically the fact that the former entirely contains the latter within its borders. *The nervous system, including its natural core, the brain, is located in its entirety within the territory of the body proper and is fully conversant with it.* As a consequence, body and nervous system can *interact directly and abundantly.* Nothing comparable holds for the relation between the world external to our organisms and our nervous systems. An astonishing consequence of this peculiar arrangement is that feelings are not conventional perceptions of the body but rather *hybrids,* at home in both body and brain.

This hybrid condition may help explain *why there is a profound distinction but no opposition between feeling and reason,* why we are *feeling creatures that think and thinking creatures that feel.* We go through life feeling or reasoning or both, as required by the circumstances. Human nature benefits from an abundance of explicit and non-explicit types of intelligence and from the use of feeling and reason, each alone or in combination.

Plenty of intellectual power, obviously, though not nearly enough for us to behave decently to our fellow humans, not to mention other living creatures.

Armed with important new facts, we are finally prepared to address consciousness directly. *How does the brain provide us with mental experiences that we unequivocally relate to our beings—to ourselves?* The possible answers, as we shall see, become disarmingly transparent.

2

Before we proceed, I need to say a few words about how I approach the investigation of mental phenomena. To be sure, the approach begins with the mental phenomena themselves, when singular individuals engage in introspection and report on their observations. Introspection has its limits, but it has no rival, let alone a substitute. It provides the only direct window into the phenomena we wish to understand, and it memorably served the scientific and artistic genius of William James, Sigmund Freud, Marcel Proust, and Virginia Woolf. More than one century later, we can claim some advances but their achievement remains extraordinary.

The results of introspection can now be connected and enriched by results obtained with other methods that also concern mental phenomena but investigate them obliquely by focusing on their (a) behavioral manifestations and (b) biological, neurophysiological, physicochemical, and social correlates. In recent decades several technical advances have revolutionized these methods and given them considerable power. The text you are about to read relies on results culled from integrating such formal scientific efforts with the results of introspection.

There is little merit in complaining about the flaws of self-observation and about its obvious limits or, for that matter, in complaining about the indirect nature of the sciences that concern mental phenomena. There is no other way of proceeding, and the multifaceted techniques that have become state of the art go a long way toward minimizing the difficulties.

One last word of caution: the facts generated by this multipronged approach require interpretation. They suggest ideas and theories meant to explain facts in the best possible way. Some ideas and theories fit the facts quite well and are rather convincing, but make no mistake: they, in turn, need to be treated as hypotheses, put to appropriate experimental tests, and supported, or not, by evidence. We should not confuse theory, no matter how seductive, with verified facts. On the other hand, it is also the case that in discussing phenomena as complex as mental events are, we often have to settle for plausibility when verification is nowhere near.

I

On Being

IN THE BEGINNING WAS NOT THE WORD

In the beginning was not the word; that much is clear. Not that the universe of the living was ever simple, quite the contrary. It was complex from its inception, four billion years ago. Life sailed forth without words or thoughts, without feelings or reasons, devoid of minds or consciousness. And yet living organisms sensed others like them and sensed their environments. By sensing I mean the detection of a "presence"—of another whole organism, of a molecule located on the surface of another organism or of a molecule secreted by another organism. Sensing is *not* perceiving, and it is *not* constructing a "pattern" based on something else to create a "representation" of that something else and produce an "image" in mind. On the other hand, sensing is the most elementary variety of cognition.

Even more surprising, living organisms responded *intelligently* to what they sensed. Responding with intelligence meant that the response helped the continuation of their life. For example, if what they sensed posed a problem, an intelligent response was one that solved the problem. Importantly, however, the smartness of these simple organisms did not rely on explicit knowledge of the sort our minds use today, the sort that requires representations and images. It relied on a concealed competence that took into account the goal of maintaining life and nothing but. This non-explicit intelligence was in charge of curating life, managing it in accordance with the rules and regulations of *homeostasis*. Homeostasis? Think of homeostasis as a collection of how-to rules, relentlessly executed according to an unusual manual of directions without any words or illustrations. The directions ensured that the parameters on which life depended—for example, the presence of nutrients, certain levels of temperature or pH—were maintained within optimal ranges.

Remember: in the beginning no words were spoken and no words were written, not even in the exacting manual of life regulations.

THE PURPOSE OF LIFE

I know that talking about the purpose of life can cause some discomfort, but considered from the innocent perspective of each living organism, life is inseparable from one apparent goal: its own maintenance, for as long as death from aging does not come calling.

Life's most direct path to achieving its own maintenance is by following the dictates of homeostasis, the intricate set of regulatory procedures that made life possible when it first bloomed in early single-cell organisms. Eventually, when multicellular and multisystem organisms became all the rage—that was about three and a half billion years later—homeostasis was assisted by newly evolved coordinating devices known as nervous systems. The stage was set for those nervous systems to not just manage actions but also represent patterns. Maps and images were on their way, and minds—the feeling and conscious minds that nervous systems made possible—became the result. Gradually, over a few hundred million years, homeostasis began to be partly governed by minds. All that was needed now for life to be managed even better, was creative reasoning based on memorized knowledge. Feelings, on the one hand, and creative reasoning, on the other, came to play important parts in the new level of governance that consciousness allowed. The developments amplified the purpose of life: survival, to be sure, but with an abundance of well-being derived in good part from the experience of its own intelligent creations.

The goal of survival and the dictates of homeostasis are still at work today, both in single-cell creatures such as bacteria and in ourselves. But the kind of intelligence that assists the process is different in single cells and in humans. Non-explicit, non-conscious intelligence is all that the simpler and mindless organisms have available. Their intelligence lacks the riches and the power generated by overt representations. Humans have both kinds of intelligence.

As we discuss life and the kinds of intelligent management that different species rely on, it becomes clear that we need to identify the menu of specific and distinct strategies available to those creatures and give names to the functional steps they constitute. *Sensing* (detecting) is most basic, and I believe it is present in all living forms. *Minding* is next. It requires a nervous system and the creation of representations and images, the critical component of minds. Mental images flow relentlessly in time and are infinitely open to manipulation so as to yield novel images. As we will see, minding opens the way to *feeling* and *consciousness*. There is not much hope of elucidating consciousness if we do not insist on distinguishing these intermediate steps.

THE EMBARRASSMENT
OF VIRUSES

The mention of intelligent but unminded competences makes me think of the tragedy we have been
living through and of the unanswered questions that
pertain to viruses. In spite of our success in managing polio and measles and HIV and coping with
the inconvenience and dangers of the seasonal flu,
viruses remain a major cause of scientific and medical
humiliation. We are negligent in our preparation for
viral epidemics, and we are ignorant when it comes
to the science we need in order to speak about viruses
clearly and deal with their consequences effectively.

We have made great progress in understanding
the role of bacteria in evolution and their interdependence relative to humans, which is largely ben-
eficial to us. The microbiome is now a part of how
we understand ourselves, but nothing comparable
holds for viruses. Our troubles begin with how to
classify viruses and understand their role in the gen-

eral economy of life. Are viruses alive? No, they are not. Viruses are not living organisms. But then why do we talk about "killing" viruses? What is the status of viruses in the big biological picture? Where do they fit in evolution? Why and how do they wreak havoc among real living things? The answers to these questions are often tentative and ambiguous, which is surprising given how much viruses cost in human suffering. Comparing viruses and bacteria is most informative. Viruses do not have energy metabolism, but bacteria do; viruses do not produce energy or waste, but bacteria do. Viruses cannot initiate movement. They are concoctions of nucleic acids—DNA or RNA—and some assorted proteins.

Viruses cannot reproduce on their own, but they can invade living organisms, hijack their life systems, and multiply. In brief, they are not living but can become parasitic of the living and make a "pseudo" living while, in most instances, destroying the life that allows them to continue their ambiguous existence and promoting the manufacture and dissemination of "their" nucleic acids. And on that point, in spite of their nonliving status, we cannot deny viruses some fraction of the non-explicit variety of intelligence that animates all living organisms beginning with bacteria. Viruses carry a hidden competence that manifests itself only once they reach suitable living terrain.

BRAINS AND BODIES

Any theory that bypasses the nervous system in order to account for the existence of minds and consciousness is destined to failure. The nervous system is the critical contributor to the realization of minds, consciousness, and the creative reasoning that they allow. But any theory that relies *exclusively* on the nervous system to account for minds and consciousness is also bound to fail. Unfortunately, that is the case with most theories today. The hopeless attempts to explain consciousness exclusively in terms of nervous activity are partly responsible for the idea that consciousness is an inexplicable mystery. While it is true that consciousness, as we know it, only fully emerges in organisms endowed with nervous systems, it is also true that consciousness requires abundant interactions between the central part of those systems—the brain proper—and varied non-nervous parts of the body.

What the body brings to the marriage with a nervous system is its foundational biological intelligence, the non-explicit competence that governs life as it meets homeostatic demands and that eventually is expressed in the form of feeling. The fact that, in good part, feeling is only fully realized thanks to nervous systems does not change this fundamental reality.

What nervous systems bring to the marriage with the body is the possibility of making knowledge explicit, by way of constructing the spatial patterns that, as we will clarify later, constitute *images*. Nervous systems also help commit to memory the knowledge represented in images and open the way for the sort of image manipulation that enables reflection, planning, reasoning, and, ultimately, the generation of symbols and the creation of novel responses, artifacts, and ideas. The marriage of bodies and brains even manages to reveal some of the secret knowledge of biology, in other words, the rhymes and reasons of intelligent life.

NERVOUS SYSTEMS AS
AFTERTHOUGHTS OF NATURE

Nervous systems came late in the history of life. No, nervous systems were not primary on any count. Nervous systems showed up to serve life, to make life possible when the complexity of organisms required high levels of functional coordination. And yes, nervous systems helped generate remarkable phenomena and functions that were not present before their arrival such as feelings, minds, consciousness, explicit reasoning, verbal languages, and mathematics. In a curious way, these "neuro-authorized" novelties expanded the achievements of the non-explicit biological intelligences and non-explicit cognitive abilities that were already in place and that had the singular purpose of serving life. The neural novelties worked toward optimizing homeostatic regulation and maintaining life more securely. This is precisely what nervous systems have been achieving by delivering the high levels of functional coordination required by complex multicellular and multisystem organisms. Complex, multicellular organisms with differentiated systems—endocrine, respiratory, digestive, immune, reproductive—were saved by nervous systems, and organisms with nervous systems came to be saved by the things nervous systems invented—mental images, feelings, consciousness, creativity, cultures.

Nervous systems are splendid "afterthoughts" of a non-minded, non-thinking, but pioneeringly prescient nature.

ON BEING, FEELING, AND KNOWING

The history of living organisms began four billion years ago and has taken several paths. In the branch of life history that led to us, I like to imagine three distinct and consecutive evolutionary stages. A first stage is hallmarked by *being;* a second is dominated by *feeling;* and a third is defined by *knowing* in the general sense of the term. Curiously, in each contemporary human, something can be gleaned akin to those same three stages, and they develop in the same sequence. The stages of being, feeling, and knowing correspond to separable anatomical and functional systems that coexist inside each of us humans and are engaged as needed in adult life.[1]

The simplest living organisms—those with only one cell (or very few cells) and without a nervous system—are born, become adults, defend themselves, and eventually die from old age, from disease, or from being destroyed by other creatures.

They are individual beings, capable of picking the best spots in their environments to live well and capable of fighting for their lives even if they do so without the help of a mind, let alone of consciousness. They have no nervous system either. Their choices lack both premeditation and reflection; you can neither premeditate nor reflect in the absence of a mind illuminated by consciousness. These beings do what they do based largely on efficient chemical processes guided by a fine-tuned but hidden competence attuned to the dictates of homeostasis so that most parameters of the life process can be maintained at levels compatible with survival. This is achieved without the help of explicit representations of the environment or the interior—in other words, without a mind—and without the assistance of thinking and thought-based decision making. The process is complemented by a minimal form of cognition manifested, for example, as "sensing" of obstacles or estimating the number of other organisms present at a given moment in a certain space, an ability known as "quorum sensing."[2]

Hidden competences reflect physical and chemical constraints and are a means of satisfying a goal—the good life, by which I mean an efficiently regulated life, capable of surviving threats—while respecting reality. Each of these competent living organisms is, in essence, a chemical factory independently operating a metabolic business and producing metabolic goods, in spite of having no digestive or circulatory systems at all. But here is something unexpected about their affairs: such "pseudo simple" creatures, best exemplified by bacteria, can live as members of a social group out in the big, wide world, namely inside other living organisms such as ourselves. We provide room and board and charge some rent in the form of helpful chemical services. On occasion, of course, the renters abuse their situation and take more than they should from the deal, and sometimes things do not end well for landlords *and* renters.

The early stage of being does not include anything that we might call explicit feeling or explicit knowledge, although the "good life" process must conform with ideal physical arrangements without which life would not have begun or would have easily fallen apart. And so, in the broad historic path we are describing here, being is followed by feeling. And, as I see it, for creatures to be able to feel, they first need to add several features to their organisms. They must be multicellular, and they must possess differentiated organ systems, more or less elaborate, among which shines a nervous system, a natural coordinator of internal life processes and of dealings with the environment. What happens then? Plenty, as we shall see.

Nervous systems enable both complex movements and, eventually, the beginning of a real novelty: *minds*. Feelings are among the first examples of mind phenomena, and it is difficult to exaggerate their significance. Feelings allow creatures to represent in their respective minds the state of their own bodies preoccupied with regulating the internal organ functions required by the necessities of life: feeding and drinking and excreting; defensive posturing such as occurs during fear or anger, disgust or contempt; social coordination behaviors such as cooperation, conflict; the display of flourishing, joy, and exaltation; and even of those behaviors related to procreation.

Feelings provide organisms with *experiences* of their own life. Specifically, they provide the owner organism with a scaled assessment of its relative success at *living,* a natural examination grade that comes in the form of a quality—pleasant or unpleasant, light or intense. This is precious and novel information, the kind of information that organisms confined to a "being" stage cannot obtain.

Not surprisingly, feelings are important contributors to the creation of a "self,"[3] a mental process animated by the state of the organism, and are anchored in its body frame (the frame constituted by muscular and skeletal structures), and oriented by the perspective provided by sensory channels such as vision and hearing.

Once being and feeling are structured and operational, they are ready to support and extend the sapience that constitutes the third member of the trio: *knowing*.

Feeling provides us with knowledge of life in the body and, without missing a beat, makes that knowledge conscious. (In chapters III and IV we shall explain how feelings manage to do so.) This is a pivotal, fundamental process, and yet, in a most ungrateful manner, we barely notice it, distracted as we are by the thunder of another branch of knowing, the one that is constructed by the sensory systems—vision, hearing, body sensations, taste, and smell—with the help of memory. The maps and images created on the basis of sensory information become the most abundant and diverse constituents of mind, side by side with ever present and related feelings. More often than not, they dominate the mental proceedings.

Curiously, each sensory system is, in and of itself, devoid of conscious experience. The visual system, for example, our retinas, visual pathways, and visual cortices, produces maps of the outside world and contributes the respective, explicit visual images. But the visual system would not allow us to automatically declare those images as our images, as occurring *inside* our organism. We would not relate those images to our beings, we would not be conscious of those images. Only the coordinated operation of the three kinds of processing—the kinds that have to do with being, feeling, and knowing—allows the images to be connected to our organism, literally *referred to it* and *placed within it*. Only then can experience emerge.

What follows from this momentous but unheralded physiological step is nothing short of extraordinary. Once experiences begin to be committed to memory, feeling and conscious organisms are capable of maintaining a more or less exhaustive history of their lives, a history of their interactions with others and of their interaction with the environment, in brief, a history of each individual life as lived inside each individual organism, nothing less than the armature of personhood.

A CALENDAR OF LIFE

Protocells	4 billion years
First cells, (or prokaryotes, such as bacteria), without a nucleus	3.8 billion years
Photosynthesis	3.5 billion years
First single cells with a nucleus (or eukaryotes)	2 billion years
First multicellular organisms	700–600 million years
First nervous cells	500 million years
Fish	500–400 million years
Plants	470 million years
Mammals	200 million years
Primates	75 million years
Birds	60 million years
Hominids	14–12 million years
Homo sapiens	300 thousand years

II

About Minds and the New Art of Representation

INTELLIGENCE, MINDS, AND CONSCIOUSNESS

Here are three treacherous concepts, and the job of clarifying what they stand for is never finished. Intelligence, in the general perspective of all living organisms, signifies the ability to resolve successfully the problems posed by the struggle for life. There is quite a distance, however, between the intelligence of bacteria and human intelligence, a distance of billions of years of evolution to be precise. The scope of such intelligences and their respective achievements are predictably different too.

Explicit human intelligences are neither simple nor small. Explicit human intelligences require a mind and the assistance of mind-related developments: *feeling* and *consciousness*. They require *perception,* and *memory,* and *reasoning*. The contents of minds are based on *spatially mapped patterns* that represent objects and actions. To begin, the contents correspond to the objects and actions that we perceive both in the interior of our organisms and in the world around us. The contents of the spatially mapped patterns we built can be *mentally inspected.* Considering a particular pattern, we the owners of the mind can inspect the "metrics" of the pattern or its "extension." Moreover, we owners of the patterns can mentally inspect their structures, relative to a specific object, and reflect, for example, on the degree of "resemblance" to that original object.

Finally, the contents of mind are *manipulable,* meaning that we, owners of the patterns, can mentally chop them in parts and rearrange the parts in myriad ways to yield novel patterns. When we attempt to solve a problem, reasoning is the name we give to the cutting and moving about that we engage in as we pursue a solution.

A convenient way of referring to the mental patterns that constitute minds is the word *images.* By images I do not mean "visual" images only but rather *any* patterns produced by the dominant sensory channels: visual, of course; auditory; tactile; visceral. When we play creatively in our minds, we do use our *imaginations,* correct?

By contrast, the intelligence of bacteria is hidden, non-explicit. None of its machinations are transparent to the observer or—and this is most important—to the intelligent organisms themselves. All that we frustrated observers know about the solving of a problem is the beginning and the end, namely, the question and the answer. As for the organisms themselves, I believe they know even less! To the best of our knowledge there is nothing in the interior of an intelligent bacterium that could construct the patterns that represent objects or actions, in their surround or in their interior, nothing that would resemble images, and therefore nothing that could resemble reasoning. But intelligent behavior works beautifully on the basis of well-articulated bioelectrical computations whose theater of operation is small—rather than simple—and sits at molecular level and below, in the physical undergirding of a living organism.

The key descriptors of the two kinds of intelligence can now be aligned for clarity: covert, hidden, concealed, *recondite,* non-explicit intelligences, to one side; overt, manifest, explicit, mapped, and mental/minded intelligences, to the other. But different as their manners are, the two kinds of intelligences came into existence to perform the same job—solving problems posed by the struggle for life. Covert intelligences solve problems simply and economically. Explicit intelligences are complicated. They require feeling and consciousness. They have made organisms care for the struggle and, in the process, invented new means to do so.

It is easy to miss the significance of the distinctions I am drawing here between non-explicit and explicit forms of intelligence. Non-explicit does not mean "magic" although plenty of biological mysteries await elucidation. Explicit does not mean fully explained either. It is simply that non-explicit mechanisms are not transparent and inspectable without the aid of such things as microscopes or fine biochemistry, not to mention a theoretical account to make sense of the facts; on the other hand, explicit mechanisms can be largely inspected by following the trail of imagetic patterns, their actions and their relationships.

As we will discover, explicit processes require *the construction and storing of imagetic patterns by the organism and inside the organism.* Moreover, that same organism must be able to inspect the patterns internally, without the help of fancy scientific technology, and organize behaviors accordingly.

INTELLIGENCES	
·covert	·overt
·hidden, concealed	·manifest
·non-explicit	·explicit
·based on chemical/ bioelectrical processes in organelles and cell membranes	·based on spatially mapped neural patterns which "represent and resemble" objects and actions; imagetic

Bacteria and other unicellular creatures have benefited from the remarkable gift of non-explicit intelligence. We humans, on the other hand, enjoy a far greater privilege. We benefit from *both* explicit and non-explicit varieties of intelligence. We use one or the other or both as needed by the problem at hand, and we do not even have to decide on which one to use. Our mental habits and styles of mentation decide for us.[1]

I leave aside one vexing issue: the intelligence of those monstrous, nonliving concoctions known as viruses. Once viruses enter a suitable living organism and even while their status remains "nonliving," they "act" most intelligently from the point of view of their permanence. The situation, as noted earlier, is a paradox and an embarrassment that we must accept. Viruses are nonliving things that act intelligently so as to foster the expansion of their potentially life-producing cargo: nucleic acids.

SENSING IS NOT THE SAME
AS BEING CONSCIOUS AND
DOES NOT REQUIRE A MIND

All living organisms, no matter how small, have the ability to detect—or "sense"—sensory stimuli. Examples of sensory stimuli include light, heat, cold, vibration, a poke. Organisms can also respond to what is sensed, and the response is aimed at either the environment that surrounds them or the interior of their body as defined by the cellular membrane that contains it.

Bacteria are capable of sensing, and the same happens with plants, and yet to the best of our judgment neither bacteria nor plants are conscious. They sense and respond to what is sensed; their cellular membranes can detect temperature, or acidity, or a micro-push and a micro-shove, and they can respond by avoiding such stimuli or, for example, by moving away from such stimuli. Bacteria and plants have a basic form of cognition and remarkable intelligence, but they do not have *explicit* knowledge concerning the things they do, nor do they have the ability to reason explicitly. How could they? Knowledge only becomes explicit to an organism once it is expressed in the form of imagetic patterns, in a mind, and the ability to reason explicitly requires the logical manipulation of imagery. Neither bacteria nor plants appear to have a mind or to be conscious. Importantly, *neither bacteria nor plants have a nervous system.*

Sensing alone does not entitle an organism to mind or consciousness. There is a precedent to be observed, however. Consciousness only becomes possible in organisms capable of sensing and capable of making minds.

Bacteria around us and within us are endowed with a *non-explicit competence* that allows them to govern their lives not just efficiently but *intelligently.* The same happens with plants. Their intelligence concerns unstated goals, namely, survive always and flourish often. Bacteria and plants operate as they "should," according to the imperatives of life regulation (or homeostasis), but they do so *blindly*—by which I mean that they do not *know* why or how they do what they do. The chemical machinery that runs their actions so successfully is not *represented* in another part of their organisms, and it has no possibility of *revealing* itself to the owner organism. *The parts and the mechanisms involved in the organism's success or failure do their job but are never "pictured" elsewhere within that organism.* Nowhere within such organisms can the parts or the machinations ever constitute explicit knowledge.

As we discuss the mindless and non-conscious nature of sensing, we should introduce and reflect on an intriguing fact: bacteria as well as plants respond to numerous anesthetics by suspending their life activities and turning to a sort of hibernation where their ability to sense disappears. These facts were first established by no less a figure than the French biologist Claude Bernard in the late nineteenth century. Imagine the astonishment of Claude Bernard when he discovered that the early, inhalable anesthetics of his day would quiet plants down to a slumber.[1]

The fact is especially noteworthy because, as we have just noted, neither plants nor bacteria appear to have minds or consciousness, the "functions" that, to this day, most everyone, commoner or scientist, associates with the action of anesthetics. You undergo anesthesia before surgery so that the loss of "consciousness" lets the surgeon work in peace and saves you from suffering. Well, I propose that what anesthetics cause—thanks to a perturbation of ion channels in the bilayer properties of cell membranes—is a radical and basic disruption of the *sensing* functions we have just described. Anesthetics do not target minds specifically—minds are no longer possible once sensing is blocked. And anesthetics do not target consciousness either, because, as we will propose, consciousness is a particular state of mind and it cannot occur in the absence of mind.

Once we are capable of consciousness, what we become conscious of is the *contents* of our minds.

Minds equipped with feeling and with some perspective on the world around them are conscious and are widely present in the animal kingdom, not just in humans. All mammals and birds and fish are minded and conscious, and I suspect that so are social insects. But I draw the line at simpler unicellular organisms. How do they do all the smart things they do? Well, we have just seen that humble bacteria have a not-so-humble competence to run their lives. They have some precursors to what will eventually permit the development of minds and even consciousness. But bacteria are not quite ready for the big time we call mind let alone a conscious mind.

THE CONTENTS OF MINDS

Turn a mind inside out and spill its contents. What do you find? Images and more images, the sorts of images that complicated creatures, such as we are, manage to generate and combine in a forward-flowing stream. This is the very "stream" that immortalized William James and gave fame to the word "consciousness" because the two words were so often paired in the phrase "stream of consciousness." But we will see that the stream, to begin with, is simply made of images whose near-seamless flow constitutes a mind. Of course, minds do become conscious once additional ingredients come to the rescue.

The perceptions of objects and actions out in the world turn into images, thanks to sight, sound, touch, olfaction, and taste. They tend to dominate our mental states, or so it seems. A good many images in our minds, however, come not from the brain perceiving the world around it but rather from the brain conniving and commingling with the world *within* our bodies. One example: the pain you provoke when you inadvertently hammer the finger rather than the nail. Such complex images can also dominate our mental proceedings as they get incorporated in the mental flow.

The images of the interior are atypical for several reasons. The devices that make these images not only portray our visceral insides; they are hooked to them, connected to their chemistry in an intimate two-way interaction. The result is the production of *hybrids* called feelings. A normal mind is made of images, from the outside—conventional or straightforward—and from the inside: *special and hybrid*.

There are more sorts of images to contend with, however. When we recall the memories we have made of objects and actions and when we re-create the feelings that accompanied them, the remembrances and the re-creations also come in the form of images. Making memories largely consists of recording images in some coded form so that eventually we can recover something close to the original. And what about the translations we make of objects and actions and feelings in the languages we know—verbal languages predominantly but also the languages of mathematics and music? The translations also show up in imagetic form.

When we relate and combine images in our minds and transform them within our creative imaginations, we produce new images that signify ideas, concrete as well as abstract; we produce symbols; and we commit to memory a good part of all the imagetic produce. As we do so, we enlarge the archive from which we will draw plenty of future mental contents.

UNMINDED INTELLIGENCE

Unminded intelligence precedes the variety of intelligence based on minds by a few billion years. Unminded intelligence is concealed in the depths of biology, and the word "recondite" is an even better term for the process. Unminded intelligence is well hidden behind the workings of molecular pathways that accomplish smart things for living organisms and can assist nonliving vessels, such as viruses, in accomplishing their mission.

Unminded intelligence manifests itself widely, in reflexes, in habits, in emotive behaviors, in competition and cooperation among organisms. Be mindful of the mindless; their repertoire is wide. And, reader, please realize that we, lofty and *minded* humans, also benefit from unminded intelligence mechanisms at all hours of the day.

THE MAKING OF MENTAL IMAGERY

Where and how do images come into being? They do so courtesy of perception, and it is easier to address perception when we begin with the world around our organism. The neural activity patterns that correspond to our surround are first concocted by sensory organs such as our eyes, our ears, or the tactile corpuscles in our skin. The sensory organs work with the central nervous system, where nuclei in regions such as the spinal cord and the brain stem assemble the signals collected by the sensory organs. Eventually, after a few more intermediate stations, the cerebral cortices receive and organize the perceptual signals. Thanks to the pioneering work of physiologists such as David Hubel and Torsten Wiesel, we know that the result of this setup is the construction of maps of objects and of their territories, in varied sensory modalities, for example, sight, hearing, touch. The maps are the basis for the images we experience in our minds.[1]

We build maps when nerve cells (neurons) become active according to certain patterns, as a result of inputs arriving from sensory devices like the eyes or ears, within regions of the cerebral cortices in the visual, auditory, and tactile systems. The abundance of detail and the practical value of the material covered by these images explain why it tends to dominate our psychological present, in most standard circumstances. The relationship between what is mapped and the images we form is a close one. Creating maps with precision is essential, while vagueness is costly. A vague map can lead you to the wrong interpretation or worse: guide you to make the wrong movement.

The attentive reader will note that I did not mention making maps and images for taste or smell, even though both are important sensory channels; nor did I mention making maps and images of the interior, an important step in the making of feelings.

The arrangements that produce smell and taste exhibit the general logic of the three major senses but exploit their own blends of chemistry and pattern assembly. They share traits of both hidden and overt forms of intelligence, and perhaps they should be seen as transitions from one to the other.[2]

On the other hand, feelings, as we will show when we discuss affect, are thoroughly hybrid processes that depend on the unique features and design of interoception, the process that opens our interior to sensory and eventually mental inspection.

The information provided by feelings points to "qualities" of things or states—good or not so good—as well as "quantities" of those qualities: really awful versus not so bad. Precision is not at a premium, and on occasion the information that feelings provide is *intentionally* incorrect by system design. This is what happens, for example, when internally manufactured opiates reduce the acute pain of a wound without the intervention of your doctor or any drugs.

TURNING NEURAL ACTIVITY
INTO MOVEMENT AND MIND

Understanding how the firing of a neuron creates movement is no longer a mystery. First, the bioelectrical phenomena of firing neurons ignites a bioelectrical process in muscle cells; second, that process causes muscular contraction; third, as a result of muscular contraction, movement happens, in the muscles themselves and in the respective bones.[1]

How a chemical-electrical process leads to mental states follows the same general logic but is far less transparent. The neural activity related to mental states is spatially distributed over arrays of neurons in a way that naturally constitutes *patterns*. The obvious examples occur in the sensory probes of vision, hearing, touch, along with those that probe the activities in our visceral interior. The patterns correspond, in spatial terms, to the objects or actions or qualities that provoke the neural activity. They *portray* the objects and actions not only spatially but also in terms of the time the actions take to unfold. The neural activity comprehensively plots the target objects and their actions on maps. The "mapped patterns" are sketched, on the fly, in accordance with the physical details of the objects and actions present in the world surrounding our nervous systems, specifically, in the world that is offered to sensory probes such as the eyes or ears. The "images" that constitute our minds are the results of the well-regimented neural activity that transmits such patterns into the brain. In other words, neurobiological "mapped patterns" turn into the "mental events" we call images. And when these events are part of a context that includes feeling and self-perspective, then, and only then, they become *mental experiences,* which is to say that they become conscious.

Depending on one's taste, one can consider this "conversion-transformation" either a magic turn of events or a very natural phenomenon. I favor the latter, but that does not mean that the explanation is complete and that all details are transparent. As I note ahead, the "physics of the mind" call for additional explanatory efforts. This "incompleteness," however, is not to be confused with the "hard problem" of consciousness. It concerns the deep *fabric* of mind, the tessitura that undergirds maps and images and that classical physics may be insufficient to fully account for. Time will tell us how hard or soft the incompleteness will prove to be.

FABRICATING MINDS

We know that our mind is made of convoys of images of assorted kinds, succeeding each other in time, from those that give us vision and sound to those that are part of feelings. We also know that the dominant images are commonly structured in a "pattern," a spatial, geometric design where elements are laid down in two or more dimensions. This spatiality is at the heart of what a mind is. It is responsible for the *explicitness* of the mental components, the precise opposite of the non-explicit competences that assist living organisms without nervous systems, quite intelligently, and that are also helpful in complex organisms such as ours. Non-explicit competences are extraordinarily effective, but the wheels of their machinery remain unavailable to mental inspection. For example, mRNA can be precisely read out into amino acid chains and even benefit from error-correcting mechanisms. However, we cannot "mentally" inspect the transcription process. Science has revealed its details, but it remains hidden from our unaided view.

Where, then, are the explicit image patterns to be found? Classical work in neuroanatomy and neurophysiology has shown that the patterns are based on "dynamic maps." These are generated at fast speed in the cerebral cortices of the varied sensory systems, including the association cortices, as well as in brain structures below the level of the cerebral cortex such as the colliculi and the geniculate ganglia. The "patterns" organized in all of these structures correspond to objects and actions and relationships present and active outside the nervous system. One way of explaining how the patterns arise is to say that sensory probes such as the retina or the cochlea analyze objects and relationships and "mimic" or "portray" them in networks of neurons, plotted in a coordinate space, while respecting real-time sequences for the objects that move. The grid-like anatomy of all these neural structures is ideal for the purpose of activating neurons in a patterned fashion so that varied designs, in varied dimensions, can be "activated" rapidly and wiped out just as rapidly.

Given the variety of cortices available in each sensory channel, we may well ask where exactly the images are assembled and experienced. Are they in the primary cerebral cortices, and if so, in which layer or layers? Or are the images in more than one cortical region, such that the actual image experienced in mind is a composite built from several simultaneously assembled patterns?

There is no definitive answer to the question of where images are. They clearly are made in varied places at different times and with different grain. Moreover, the "where" question is connected to a related inquiry: By what additional mechanism do images *become* conscious? We will deal with this inquiry after we next address feelings, the indispensable contributors to the process of making images conscious.

Perhaps an even more enigmatic question pertains to the deeper fabric of mind, the *tessitura* issue that I mentioned earlier. To say that mind processes rely on bioelectrical events in neuron circuits is certainly correct. But can we go search *beneath* that statement? It is there, I suspect, that it may be helpful to investigate the physical structure and dynamics of neural tissues and of the non-neural surroundings in which they are embedded. In this regard, physicists such as Roger Penrose, the biologist Stuart Hameroff, and the computer scientist

Hartmut Neven have suggested that quantum-level processes operating inside cells, specifically in neurons, are important players in mental events.[1]

In their favor, recent developments in general biology suggest that sub-molecular, quantum-level events are critical to account for complex biological processes such as photosynthesis. The same applies to sonar, echolocation, and the determination of magnetic north in birds, all "mind-related" phenomena.

I note that in my perspective the above considerations apply to the fabrication of mind and only to mind. As I will show in the next chapters, explaining consciousness—explaining how to make minds conscious—does *not* require us to invoke the sub-molecular level, while explaining the *fabric of mind* may. Consciousness is a systems-level phenomenon. It calls for a rearrangement of the furniture of mind, *not* the fabrication of the individual pieces.

THE MINDS OF PLANTS AND
THE WISDOM OF PRINCE CHARLES

One has to have a soft spot for a person who talks to plants, as Prince Charles is supposed to do. One has to agree that speaking to plants implies not only a recognition of worthy forms of nonhuman life but also respect for the idea that good care, actual or poetized in the form of kind words, makes a difference in the life of nonhuman organisms, a lovely thought indeed.

I have no idea if Prince Charles actually knows something about botany in particular or about biology in general, but there is plenty of reason for him to respect and love plants. And he is in good company, none other than Claude Bernard, whom we have just met. Claude Bernard uncovered the effect of anesthetics in the life of plants, grasped the significance of life regulation back in the last quarter of the nineteenth century, and explained its necessity for maintaining the balances in the physicochemical interior of all living things, to which he gave a distinctive name, the "internal milieu." Some of his thinking was inspired by the life of plants, and it is easy to imagine him talking to them as well, although one does not need to go that far. It is enough to recognize that although the term "homeostasis" only came to be a few decades later—by the pen of the American scientist Walter Cannon—the admirable Claude Bernard, working quietly in Paris, first described the phenomenon of homeostasis and realized its importance.[1]

And what did Claude Bernard see in his plants? He saw living creatures with many cells and different kinds of tissues, managing complicated multisystem organisms quite successfully in spite of being largely encased in cellulose, being deprived of muscles and thus prevented from engaging in *obvious* motion. He saw that they were actually capable of plenty of *nonobvious, stealth* movement, with their impressive network of underground roots. And those roots, how seemingly knowledgeable they were and are, growing at their slow but inexorable pace toward the region of the underground that will provide them with most water and nutrients.

Claude Bernard also realized that water could be hoisted up aboveground, to the well-exposed tops of plants and to their leaves and flowers, thanks to an efficacious system of hydraulic circulation. And he realized that multicellular, multisystem organisms had a brilliant solution for generating movement by juxtaposing new cellular elements, one next to the other, and thus "moving" the tip of a limb by elongating the whole limb. This is something that plants do when their root system bends and grows in one particular direction, toward the place where water molecules wait in abundance. Exceptionally, plants actually move by using something akin to muscles, as is the case with the leaves of the Venus flytrap, but that is not the rule.

Claude Bernard would not have been astonished to discover what we have learned since his time: that roots of trees in forests form vast networks that contribute to a collective homeostasis.[2]

All of these wonders are performed in the absence of nervous systems but with the help of abundant sensing and non-minded intelligence. But who needs a mind when one can do so much without it? Plenty of good reason, then, for Claude Bernard to admire this family of living organisms and investigate the obeisance they manifest to the imperatives of homeostasis. Plenty of good reason for Prince Charles to honor them too with his monologues.

ALGORITHMS IN THE KITCHEN

People often speak of algorithms with reverence, with the respect appropriately owed to the sort of scientific or technical development that has changed lives. The reverence and the respect are well justified, but it is important to understand the nature of algorithms and be clear about their limits especially when we compare them to images. One should think of algorithms as recipes, as the way to prepare Wiener schnitzel or, as Michel Serres has suggested, tarte tatin.[1] Recipes are helpful, of course, but they are not the thing that the recipes are meant to help you reach. You cannot taste a recipe of Wiener schnitzel or savor a recipe for tarte tatin. Thanks to your mind, you can *anticipate* the tastes and salivate accordingly, but given a recipe alone, you cannot really savor a nonexistent product. When people think of "uploading or downloading their minds" and becoming immortal, they should realize that their adventure—in the absence of live brains in live organisms—would consist in transferring *recipes,* and only recipes, to a computer device. Following the argument to its conclusion, they would not gain access to the actual tastes and smells of the real cooking and of the real food.

I am not disparaging algorithms. How could I, after all the hymns of admiration I have sung for recondite intelligences and for the codes that enable them?

III
On Feelings

THE BEGINNINGS OF FEELING: SETTING THE STAGE

Feeling probably began its evolutionary history as a timid conversation between the chemistry of life and the early version of a nervous system within one particular organism. In creatures far simpler than we are, the exchange would have generated feelings such as plain well-being and basic discomfort rather than subtly graded feelings, let alone something as elaborate as localized pain. Still, what a remarkable advance. Those timid beginnings provided each creature with an orientation, a subtle adviser as to what to do next or not to do or where to go. Something novel and extremely valuable had emerged in the history of life: *a mental counterpart to a physical organism.*[1]

AFFECT

The simplest variety of affect begins in the interior of a living organism. It springs up vague and diffuse, generating feelings that are not easily described or placed. The term "primordial feelings" captures the idea.[1] By contrast, "mature feelings" provide vivid and assertive images of the objects that furnish our "interior"—viscera such as the heart and lungs and gut—and of the actions they execute such as pulsing and breathing and contracting. Eventually, as in the case of localized pain, the images become sharp and focused. But make no mistake: vague, approximate, or precise, feelings are *informative;* they carry important knowledge and plant that knowledge firmly within the mind flow. Are muscles tense or relaxed? Is the stomach full or empty? Is the heart beating regularly and boringly, or is it skipping beats? Is the breathing easy or labored? Is there pain in my shoulder? We, who have the privilege of feeling, get to know about such states, and that information is valuable for the subsequent governance of our lives. But how do we come by such knowledge? What happens when we "feel," as opposed to when we simply "perceive" objects in the world at large? What is required for us to *feel,* as opposed to merely perceive?

First, *everything we feel corresponds to states of our interior.* We do not "feel" the furniture around us or the landscape. We can perceive the landscape and the furniture, and our perceptions can easily elicit emotive responses and result in the respective feelings. We can *experience* these "emotive feelings" and even name them—the *beautiful* landscape and the *pleasant* chair.

But what we "really" feel, in the proper sense of the term, is how either parts or the whole of our own organism are faring, moment by moment. Are their operations smooth and unimpeded, or are they labored? I call these feelings homeostatic because, as direct informers, they tell us if the organism is or is not operating according to homeostatic needs, that is, in a manner conducive or not to life and survival.

Feelings owe their existence to the fact that the nervous system has direct contact with our insides and vice versa. The nervous system literally "touches" the organism's interior, everywhere in that interior, *and* it is "touched" in return. The nakedness of the interior relative to the nervous system and the direct access the nervous system enjoys relative to that interior are part of the uniqueness of interoception, the technical term reserved for the perception of our visceral interior. Interoception is distinct from the perception of our musculoskeletal system, known as *proprioception,* and from the perception of the outside world, or *exteroception.* We can obviously use words to describe the experiences of feeling, but we do not need the mediation of words in order to feel.[2]

Feelings, as enacted in our organism and experienced in our minds, exert a tug and a pull over us, literally disturb us, positively or negatively. Why and how can they do so? The first reason is clear: they are "insiders," and they have access to our interior! The neural machinery that helps us "manufacture a feeling" interacts directly with the object that caused the feeling. For example, pain signals hailing from the capsule of a sick kidney travel into the central nervous system and coalesce to become a "renal colic." But the process does not stop there. The central nervous system engenders a response back to the sick kidney's capsule and modulates the continuation of the pain; it may even interrupt it. Other events in the area—for example, local inflammation—generate their own signals and contribute to the experience. The overall situation claims one's attention and involvement.

The example of the renal colic we just considered helps us illustrate the point that feelings are assembled by an elaborate physiology distinct from the physiology the organism uses for vision or hearing. Rather than pinpointing a particular exterior feature such as one particular shape or sound with precision and stability, feelings often correspond to a range of possibilities. Feelings depict certain *qualities* within a scale and their *variations* in terms of tone and intensity. Figuratively, feelings do not take simple snapshots of external objects or events; feelings tape the whole show and the backstage activity, not just the surfaces, but also what is underneath.

Feelings are *interactive perceptions*. Compared with visual perceptions—the canonical example of perception—feelings are *unconventional*. Feelings gather their signals "inside the organism" and even "inside the objects located in that inside" rather than simply around the organism. Feelings depict actions that occur in our interior, as well as their consequences, and let us catch a glimpse of the viscera involved in those actions. Little wonder that feelings exert a special power over us.

The operations of interior organs and systems are

gradually represented in the nervous system, first in its peripheral nerve components, then in nuclei of the central nervous system (in the brain stem, for example), and later in the cerebral cortex. But there is an intense cooperation between body parts and neural elements. Body and nervous system remain interactive partners rather than separate "model" and "depiction." What is ultimately imaged is neither purely neural nor purely bodily. It emerges from a dialogue, from a dynamic give-and-take between body chemistry and the bioelectrical activity of neurons. And, to make matters more complicated, at any moment an emotive response, such as fear or joy, can impose further changes in some viscera—which are the primary body actors in the emotive process—and generate, as a result, a new set of visceral states and a new set of brain-body partnerships. Such emotive responses change the organism and consequently change what is to be imaged via the body-brain partnership. The result is a new set of feelings—now partly "emotional" rather than purely "homeostatic"—and a new affective state. Moods are the consequence of this kind of dynamic, held over long periods of time. They are the origin of the "enthusiasm" or "lassitude" with which we enter each new day. So are varied degrees of excitement/arousal and dullness/sleepiness.

. . .

The following definitions should make the above descriptions even clearer.

Homeostasis: as we saw earlier, homeostasis is the process of maintaining the physiological parameters of a living organism (for example, temperature, pH, nutrient levels, visceral operations) within the range most conducive to optimal function and survival. (The related but distinct term "allostasis" refers to the mechanisms used by an organism as it seeks to regain homeostasis.)[3]

Emotions: collections of co-occurring and involuntary internal actions (for example, smooth muscle contractions, changes in heart rate, breathing, hormonal secretions, facial expressions, posture) triggered by perceptual events. The emotive actions are usually aimed at supporting homeostasis, for instance, countering threats (with fear or anger) or signaling successful states (with joy). When we recall events from memory, we also produce emotions.

Feelings: the mental experiences that follow and accompany varied states of organism homeostasis, whether primary (*homeostatic feelings* such as hunger and thirst, pain or pleasure) or provoked by emotions (*emotional feelings* such as fear, anger, and joy).[4]

No matter what the "precise" contents of your mind may be—the landscapes, the furniture, the sounds, the ideas—those contents are necessarily experienced *together with affect*. What you perceive or remember, what you try to figure out by reasoning, what you invent or wish to communicate, the actions you undertake, the things you learn and recall, the mental universe made up by objects, actions, and abstractions thereof, *all* of these different processes *can generate affective responses as they unfold*. We can think of affect as the universe of our ideas transmuted in feeling, and it is also helpful to think of feelings in music terms. Feelings perform the equivalent of a musical score that accompanies our thoughts and actions.

The non-feeling, "precise" contents of the mind flow with distinction, silhouetted against the affect process, a bit like acting figurines against an animated backdrop. But these precise contents often interact with the process of affect. At any moment, one actor or actors within the "precision content" troupe may succeed in stealing the show and making it "be" different by provoking new emotions and producing the corresponding feelings. Some interesting variations on the musical score that is being improvised will follow, in good order. To make matters really fascinating, the opposite is also true: affect can alter the lights under which the precision contents are experienced. Affect can alter how long the images stay on the mind's stage and how well or not so well they are perceived. Precise contents, on the one hand, and affect, on the other, are distinct in terms of how organisms construct them, and are fully interactive. We should celebrate the wealth and the messiness we have been gifted by affect.

BIOLOGICAL EFFICIENCY AND THE ORIGIN OF FEELINGS

The notion of efficiency sounds like a human invention meant to describe the modern world, but it applies simply and well to the early life of billions of years ago and to its successful operations in terms of energy consumption. Efficiency was regimented by homeostasis and made even more successful by natural selection. How the degree of obeisance to homeostasis results in greater or lesser energy consumption is an old life trick, not a new development. Bacteria have been exploiting efficiencies quite well, and so have numerous mindless but successful species, between bacteria and humans.

How intriguing, then, that in the course of natural history, feeling became a part-time guide to good governance. How did *that* happen? A starting point must have been the alignment of efficiency and survival with certain parameters of physics and chemistry, while dysfunction and death aligned with certain other parameters. There is nothing wrong with the idea of a Platonic "form of the good" that would be present—almost certainly is—in the physics that undergird life and thriving.[1] But as I see it, the remarkable expansion and promotion of one choice—the life-favoring arrangements—over the pain and suffering alternative, came courtesy of feelings which really means courtesy of consciousness. *All* feelings are conscious, and while disagreeable feelings signify situations that impede and endanger life, pleasant ones signify those that help life flourish. In the absence of feelings/consciousness, the mechanisms aligned with flourishing would not have gained favor so overwhelmingly. The presence of consciousness changed matters radically. Only a devil might have altered the preference that conscious feelings pointed to so clearly.

The alignment of homeostasis, efficiency, and varieties of well-being was signed in heaven, in the language of feeling, and it was made popular by natural selection. Nervous systems officiated.

GROUNDING FEELINGS I

The feelings that we humans experience could only have begun in earnest after the evolutionary rise of complex nervous systems capable of making detailed sensory mappings and images. The resulting primordial feelings were important stepping-stones on the way to the elaborate feelings humans can experience today.

The sensory maps and images that are part of elaborate feelings incorporate in the ongoing mental flow facts regarding the state of the organism's interior. This informational role is a primary contribution of feelings, but feelings have another role to play: they provide the urge and the incentive to behave according to the information they carry and do what is most appropriate for the current situation, be it running for cover or hugging the person you have missed.

GROUNDING FEELINGS II

The spontaneous chemical activity within the organism's interior is aimed at regulating life according to homeostatic dictates. The activity naturally tends toward achieving ranges of operation compatible with survival and positive energy balances, but the degree to which it succeeds varies with the organism and the situation. As a consequence the profiles of chemical activity within a particular organism correspond to—and thus stand for—degrees of success or failure in the attempt to secure homeostasis and survival. These profiles constitute a natural evaluation of the ongoing life process.

Feelings enter this picture because there is a manifest and principled correspondence between "degrees" of life-regulatory success or failure and the variety of positive or negative feelings we experience. The affective component of our mental experiences reflects the profiles of our biological processes.

The earliest physiological source of feelings is an integrated chemical profile of the organism's interior. It is likely that such a molecular-level source was present in evolution prior to the appearance of nervous systems. But this is not to say that simple organisms devoid of nervous systems would have been (or are) capable of mental experiences, beginning with the experience of feelings. Feelings reflect a chemical regulatory process, the *initial* condition without which they could not occur, but another condition must be met, and that is a dialogue between body chemistry and the bioelectrical activity of neurons in a nervous system. Regulatory chemical molecules ignite the feeling process but cannot complete it alone.

GROUNDING FEELINGS III

Perhaps we are now ready to take the Orphic plunge and descend into the feeling underworld. I have suggested that feelings originate in our deep organismic chemistry, but can we say something about how and where?

The deeper levels of the feeling process concern the chemical machinery responsible for the entire scope of homeostatic regulation along varied pathways. Underneath the qualities and intensities that constitute the valuations expressed in feelings—their valences—there are molecules, receptors, and actions.

How this chemical orchestra does its job is a bit of a marvel. Specific molecules act on specific receptors and cause specific actions. These actions are part of the uphill struggle for the maintenance of life. The actions themselves are important enough, but so is the overall dynamic of which they are a part and which is charged with managing the life

of a specific organism. This much is easy to understand. But what is not so transparent is how the
actions that result from molecules and receptors
doing their job can help us account, in our subjective experiences, for the "stirrings" that feelings
cause in us, let alone for the "quality" of a feeling.

As we try to answer the above questions, it is
helpful to recall that whereas plain percepts of
objects or actions in the world exterior to us arise
from neural probes located at the periphery of the
organism, feelings arise from the depths of our interior and not necessarily from one region only. The
retinal maps that help us see, or the skin corpuscles
that help us touch, accomplish miracles of detection
and description, but they are aloof devices relative
to our lives. They are not immediately engaged with
the miseries and glories of our life maintenance,
while feelings are.

Because the actual *object* of the feeling/percep-
tion is none other than a part of the organism itself,
that object is in fact located *within* the *subject/per-
ceiver*. Astonishing! Nothing comparable occurs
with our external perceptions, for example, visual
or auditory. The objects of visual or auditory perceptions do not communicate with our bodies. The
landscape we see or the songs we hear are *not* in

touch with our bodies, let alone inside them. They exist in a physically separate space.

In the feeling realm the situation is radically different. Because the object and subject of our feeling-percepts exist within the same organism, *they can interact.* The central nervous system can modify the body state that gives rise to a particular feeling and, by doing so, modify what is felt. *This is an extraordinary setup that has no counterpart in the world of external perceptions.* You may well want to modify an object that you are in the process of seeing; you may even wish to beautify a particular image that you are contemplating. Alas, you will not be able to *actually* do so except in your imagination.[1]

The physical disturbance that distinguishes feelings is explained by the incessant provocation of actions in the interior of our bodies, by the subsequent reflection of those actions in extensive and multiple-level neural mappings of that same interior, and by the fact that those mappings are tied to varied body compartments and actions. These mappings are the primary source of the varied "coloring" of feelings. They generate the *valences*—positive and negative, pleasurable or uncomfortable, agreeable or disagreeable—that the organism gets to experience.

The actions that arise from the body are quite varied. There can be ease and relaxation of muscle fibers, or contraction and strangulation of a particular organ, or actual movement of an internal or skeletal part. As reflected in sequential and ever more differentiated maps, the overall profiles of ease and relaxation contribute to feelings that we designate by such terms as *well-being* and *pleasure;* the contraction and strangulation patterns produce what we call *discomfort* and *malaise*. Eventually, given the detailed and interactive map of a locally strained muscle or a wound, we produce the extreme discomfort that we designate as *pain*.

The pleasure and pain felt in a particular organism begin deeper than organs and muscles. They begin with the molecules and receptors whose actions transform tissues and organs and systems in a particular organism. They continue where some of those molecules act on the neural networks that process the signals generated by the body.

GROUNDING FEELINGS IV

We have just seen how the nervous system is *inside* the body and how body and nervous system interact directly, no intermediary needed. On the other hand, the nervous system is *separate* from the world external to the organism; it maps the external world via sensory processes such as vision and hearing, which are firmly planted in the body and use the body as an intermediary.

When we say that we "represent" or "map" objects in the world around us, the notion of "mapping" introduces distance between the "map" and "the things mapped," as it should. There is often an abyss between the map and the object, as when, a few minutes ago, I went out on the terrace and watched the sun set behind the Santa Monica Mountains and saw the red twilight that followed.

We must be careful when we use the notion of mapping in relation to our own body and to the making of feelings, as if the maps were a pure

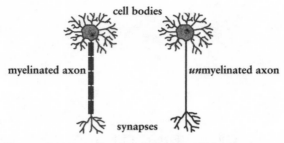

Figure III.1: Myelinated and unmyelinated axons.
The unmyelinated axons are not insulated.

"reflection" or "picture" of the body structure and
state, yet another example of a detached percept.
Our feelings are not detached at all. In practice,
there is little distance between feelings and the
things felt. Feelings are commingled with the things
and events we feel thanks to the exceptional and
intimate cross talk between body structures and
nervous system. This intimacy, in turn, is itself a
product of the peculiarities of the system charged
with signaling from the body into the nervous system, that is, the *interoceptive system*.[1]

The first peculiarity of interoception is a per-
vasive lack of myelin insulation in a majority of
interoceptive neurons. Typical neurons have a *cell
body* and an *axon,* the latter being the "cable" that
leads to the *synapse.* In turn the synapse makes
contact with the next neuron and either permits or

withholds its activity. The result is the firing of the neuron or its silence.

Myelin serves as an insulator of the axon cable, preventing extraneous chemical and bioelectrical contacts. In the absence of myelin, however, molecules in the surround of an axon interact with it and alter its firing potential. Moreover, other neurons can make synaptic contacts along the axon rather than at the neuron's synapse, giving rise to what is known as *non-synaptic signaling*. These operations are *neurally impure;* they are not really separate from the body that hosts them. By contrast, a predominance of myelinated axons insulates neurons and their networks from the influences of their surrounding environment.

A second peculiarity of interoception concerns a lack of the barrier that normally separates neural affairs from the bloodstream. This is known as the blood-brain barrier (in relation to the central ner-

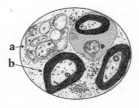

Figure III.2: Cross section through a major nerve showing (a) unmyelinated and (b) myelinated axons.

vous system) or the blood-*nerve* barrier (in the case of the peripheral nervous system). The absence of a barrier is especially notable in brain regions related to the interoceptive process, such as the spinal and brain stem ganglia where circulating molecules can make direct contact with the cell bodies of neurons.

The consequences of these peculiarities are remarkable. Lack of myelin insulation and lack of blood-brain barrier allow *signals from the body to interact with neural signals directly*. In no way can interoception be regarded as a plain perceptual representation of the body inside the nervous system. There is, rather, an extensive commingling of signals.

GROUNDING FEELINGS V

By now we should be clear about the origin of feelings. Feelings arise in the interior of organisms, in the depth of viscera and fluids where the chemistry responsible for life in all its aspects reigns supreme. I am talking about the operations of the endocrine and immune and circulatory systems, in charge of metabolism and defense.

What about the "function" of feelings? Although the history of cultures and even the history of science have made the role of feelings seem not just mysterious but unfathomable, the answer is apparent: feelings help with the management of life. More specifically, feelings operate as alerting sentinels. They *inform each mind*—fortunate enough to be so equipped—*of the state of life within the organism to which that mind belongs.* Moreover, *feelings give that mind an incentive to act according to the positive or negative signal of their messages.*

Feelings collect information about the state of life within the organism, and the "qualities and intensities" that are manifested by feelings constitute *valuations* of the process of managing life. They are direct expressions of the degree of success or failure of the life enterprise within our body. Keeping alive is an uphill battle, and our bodies engage in a complicated and multicentric effort to make life not only possible but robustly so. The robustness of life is felt as "plenitude" and "flourishing"; a balanced life process is translated as "well-being." "Discomfort" and "malaise" and "pain," on the other hand, signify failure at the life management effort.

The dramatic situation that we living creatures face concerns the maintenance of coherence and cohesion in our living organisms. The coherence and cohesion of the inanimate objects that surround me at this moment are no problem at all for those objects or for me. The objects are largely perpetual unless I decide to take an ax to the desk where I am writing, or to the chair where I sit, or to the shelves and books that surround me. Not so with my life and with the organism it animates. I need to feed them breakfast and lunch, I need to keep the body in a temperate environment, I need to prevent or avoid disease or treat it once I acquire it. I even need to maintain and nourish healthy social relationships with those around me so that circumstances arising in the social world do not impinge on the state of my interior and disturb the process of governing life in terms of homeostatic necessities.[1]

Arising as they do in the interior of our adjustable and dynamic organisms, feelings are both *qualitative and quantitative*. They exhibit *valence*—the quality rankings that make their warnings and advice be worth the effort and also motivate our actions as needed. When I experience homeostatic feelings—a situation that reflects an appraisal of my interior when certain physiological profiles prevail—I get to know, firsthand, about the state of my life, *and* the negative or positive valence of the experience advises me to correct the situation or else accept it and do little or nothing. It makes me spring into action or enjoy the ride.

Consider how different the situation is when I look at the objects around me, or hear ambient sounds, or touch an object, or see other living organisms. In that situation I am also the recipient of information. I am still being "informed" of the presence and characteristics of the objects or organisms but now the source of the data is the external world and its objects and creatures. I am being told of externalities; I am not told about the interior of the entities I see or hear or touch. A perceptual distance separates me from those entities. They *are not within my organism.*

GROUNDING FEELINGS VI

Feelings such as hunger and thirst signify quite transparently a drop in energy sources or a decline of the ideal amount of water molecules. Thankfully, given that neither reduction is compatible with the continuation of life, let alone with healthy life, feelings do more than provide valuable information: they force us to act according to the information. They motivate our actions.

The trajectory behind the process of feeling is clear: a multitude of basic micro-messages travel from body tissues and organs either to (a) circulating blood and from there to the nervous system or, directly, to (b) nerve terminals embedded in body tissues and organs. Once the signals arrive in the central nervous system—in the spinal cord and the brain stem, for example—they face a number of possible roads that lead to varied neural centers where the feeling process can be advanced further. Ultimately, such complicated signal trajectories result in the production of informative mental images. The images, such as, for example, a dry mouth, a growling stomach, or the mere lack of energy signaled by weakness, operate as indicators of trouble. They are accompanied by worry and discomfort—an emotive state—which in turn motivate a response, in the form of a corrective action.

Many of the responses that feelings promote or demand are executed automatically without any need for reasoned intervention. The extreme example to which I alluded earlier can be found in the processes of breathing and micturition. A reduction or interruption of airflow, which occurs in severe asthma or in pneumonia, is automatically accompanied by a desperate state of "air-hunger"—a literal and precise term—and the alarm it causes in the victim and in those who witness it. The need for micturition resulting from a full bladder is less dramatic than air-hunger and can even be a source of comedy but is another example of homeostatic crisis translated in forceful emotive terms and felt as an imperative, unavoidable urge.[1]

In brief, nature has provided us with the fire alarms, the fire engines, and the medical facilities. A sign that nature has been perfecting this strategy is shown by the recent discovery of central nervous system controls of immune responses. The controls are located in the diencephalon, a sector of the central nervous system located below the cerebral cortex and above the brain stem and spinal cord. The region in charge of this immune control is known as the hypothalamus, a famed orchestrator of the endocrine system that governs the secretion of most hormones throughout the body. The new findings show that the hypothalamus commands the spleen to produce antibodies to certain infective agents. In other words, the immune system works with the complicity of the nervous system to promote homeostasis without asking us, the presumed conscious controllers of our destinies, any help with the matter.

Equally intriguing is the connection between the highest neural instances of the feeling process—the insular cortices—and the innervation of the stomach mucosa. We know that stomach ulcers are directly caused by a specific bacterium, but the regulation of one's emotions is a factor in the process of allowing the bacterium to give us an ulcer or not.

GROUNDING FEELINGS VII

When we ask ourselves where homeostatic feelings begin, a reasonable first answer is that they begin with sets of molecules that signify advantageous or disadvantageous life states relative to such physiological parameters as (a) positive or negative energy balance; (b) presence or absence of (i) inflammation, (ii) infection, (iii) immune reactions; and (c) harmony or discord in the discharge of drives and goals.

The range of the critical molecules involved is wide. It includes opioids, serotonin, dopamine, epinephrine and nor-epinephrine, and substance P, all of which have a large share of operations in this domain. Some of those molecules, which are historically almost as old as life and operate on many organisms *without* nervous systems, are unfortunately known as "neurotransmitters." The misnomer is due to the fact that they were first described in creatures with brains. But the effect of these molecules does not necessarily end once they are released. The changes they impose on the operation of body systems can subsequently be translated by interoception made to influence the central nervous system and, once again, alter the mental experiences of the moment. This process is achieved via nerve fiber terminals strewn about body tissues—skin, thoracic and abdominal viscera, blood vessels—and via the projection of those nerve terminals into the spinal and the trigeminal ganglia and the spinal cord. From there neurons can signal to brain stem nuclei (the parabrachial nucleus, and the periaqueductal gray), to the amygdala nuclei, and to the nuclei of the basal forebrain. Eventually signals can reach the cerebral cortices of the insula and cingulate regions.

Not all homeostatic feelings are harbingers of bad news or signify danger ahead. When the organism is functioning with a good balance between what it requires to operate well and what it gets, when the environment is suitable in terms of climate, and when we are at ease in our social environment rather than in conflict, then the star homeostatic feeling is *well-being,* available in various guises and intensities. Well-being can become so abundant and focused that it rises to the experience of pleasure. Likewise, in the world of negative homeostatic feelings, malaise can be so acutely focused that it becomes *pain*.

The homeostatic feeling of pain offers an automatic diagnosis: damage has already occurred in some region of living tissue or is about to occur and will occur if the situation is not remedied fast. The insult must be removed or mitigated. Substance P is a critical actor in the pain process, and the secretion of cortisol and corticosterone is part of the response to the insults that lead to pain.[1]

HOMEOSTATIC FEELINGS IN A SOCIOCULTURAL SETTING

We are quite familiar with the direct way in which illness gives way to discomfort and pain or exuberant health produces pleasure. But we often overlook the fact that psychological and sociocultural situations also gain access to the machinery of homeostasis in such a way that they too result in pain or pleasure, malaise or well-being. In its unerring push for economy, nature did not bother to create new devices to handle the goodness or badness of our personal psychology or social condition. It makes do with the same mechanisms. Playwrights and novelists and philosophers have long known this fact, but it remains unappreciated perhaps because the way things work tends to be even more nebulous when it comes to society and culture than when we deal with the rigors of the medical setting.

Still, the pain of social shame is comparable to that of a raging cancer, betrayal can feel like a stab wound, and the pleasures that result from social admiration, for better and worse, can be truly orgasmic.[1]

BUT THIS FEELING
ISN'T PURELY MENTAL

The above verse appears in the song "I Won't Dance" written by Jerome Kern and made famous by Fred Astaire, Frank Sinatra, and Ella Fitzgerald. A good part of its success comes from the lyrics that Dorothy Fields and Jimmy McHugh included in the song's revised version. "But this feeling isn't purely mental" is followed by "For heaven rest us, I'm not asbestos." The naughty implication is that love is not just in the mind but also in the physical excitement that the hero notices when he dances with his beloved. He is not made of asbestos; he is a flesh-and-blood human being, and he reacts *physically* to the closeness and the romance! He is embarrassed, and he won't dance anymore.

Sometimes popular wisdom beats laborious science. That feelings are not purely mental; that they are hybrids of mind and body; that they move with ease from mind to body and back again; and that they disturb the mental peace, those are the points of the song and my points in this chapter. All I need to add is that the power of feelings comes from the fact that they are present in the *conscious mind*: technically speaking, we feel because the mind is conscious, and we are conscious because there are feelings! I am not playing with words; I am merely stating the seemingly paradoxical but very real facts. Feelings were and are the beginning of an adventure called consciousness.

IV

On Consciousness
and Knowing

WHY CONSCIOUSNESS? WHY NOW?

You may wonder why so many philosophers and scientists are writing about consciousness these days, why a topic that until recently was not prominent in the scientific literature, let alone with the public at large, is now a leading theme of scholarship and an object of curiosity. But the answer is simple: consciousness matters and the public has come to realize it.

The significance of consciousness comes from what it brings directly to the human mind and from what it subsequently allows the mind to discover. Consciousness makes mental experiences possible, from pleasure to pain, along with all that we perceive and memorize and recall and manipulate as we describe the world around and the world inside, in the process of observing, thinking, and reasoning. If we were to remove the conscious component from our ongoing mental states, you and I would still have images flowing in our minds, but those images would be unconnected to us as singular individuals. The images would not be owned by you or me or anyone else. They would flow unmoored. No one would know to whom such images belonged. Sisyphus would be fine. He is a tragic figure only because he knows that the abominable predicament is *his*.

Nothing can be *known* in the absence of consciousness. Consciousness was indispensable for the rise of human cultures and thus had a hand in changing the course of human history. It is difficult to overstate the importance of consciousness. At the same time it is easy to exaggerate the difficulty of understanding how consciousness arises and to promote it as an inscrutable mystery.

Now, why do I write about the *human* significance of consciousness, given that in all likelihood vertebrate creatures and many invertebrate species are also endowed with consciousness? Is consciousness not significant for them too? Well, it certainly is, and I am not neglecting the abilities or relevance of nonhumans. I am simply giving pride of place to the following facts: (1) the human experience of pain and suffering has been responsible for extraordinary creativity, focused and obsessive, responsible for inventing all kinds of instruments capable of countering the negative feelings that initiated the creative cycle; (2) conscious well-being and pleasure have motivated countless ways with which humans secured and enhanced conditions favorable to their lives, individually and socially. Nonhumans, with rare but notable exceptions, have also responded to pain or to well-being along the same lines but in simpler and more direct ways than humans. To be sure, nonhumans have succeeded in evading or mitigating causes of pain and suffering but, for example, have not been able to modify their origins. The consequences of consciousness for humans have been remarkably larger in scope and reach. Note that this is not because the core mechanisms of consciousness are different in humans—I believe they are not—but because the intellectual resources of humans are so much taller and wider. Those larger resources have enabled humans to respond to the polar experiences of suffering or of pleasure by inventing new objects, actions, and ideas, which have translated into the creation of cultures.[1]

There are some seeming exceptions to this panorama. A small fraction of insects, known as "social," has succeeded in assembling a complex set of "creative" responses whose ensemble does conform to the general concept of "cultures." This is the case with bees and ants and the well-organized urbanity and civility of their carefully built "cities." Are they too small and modest to have been endowed with consciousness and to have their creativity fueled by consciousness? Not at all. I suspect that they are driven by the conscious feelings they experience. The inflexibility of most of their behaviors limits the evolution of such cultural feats—a polite way of saying that they are largely "fixed" rather than evolving. But that should not diminish our amazement at how these developments came to pass a hundred thousand years ago and at the role consciousness probably played in them.

One other partial qualification regarding the special impact of consciousness in humans concerns the way in which certain mammals respond to the death of others, as is clear, for example, from the funeral rites of elephants. No doubt, consciousness of their own suffering caused by observing the results of pain and death in their kin worked its way into the composition of such responses. The difference, relative to humans, sits with the scale of invention and the degree of complexity and efficacy shown in the construction of responses. These exceptions generally support the idea that the differences of response are related to the intellectual caliber of the species rather than to the nature of consciousness in the particular species.

It is reasonable to ask if the efficacy of the responses that consciousness makes possible comes mostly from the negative or the positive side of feelings, from their negative or positive valence. Pain, suffering, and the realization of death are especially empowering, more so I believe than well-being and pleasure. In this regard, I suspect that religions developed around that realization, none more so than the Abrahamic religions and Buddhism. To some extent, in its historical, evolutionary terms, consciousness was *a* forbidden fruit that once eaten made one vulnerable to pain and suffering and ultimately exposed to a tragic confrontation with death. This perspective is closely compatible with the idea that consciousness is introduced in evolution by the hand of feelings and not just any feelings but negative ones, in particular.

Death as a source of tragedy was well established in biblical narratives and in Greek theater, and has remained present in artistic endeavors. W. H. Auden captures the idea in a poem in which he turns humans into exhausted but rebellious gladiators pleading with a cruel emperor and says, "We who must die demand a miracle." He wrote *demand* and not *require* or *request,* a sure sign of a poet at the end of his rope, watching in desperation the inescapable crumbling of the individual human. Auden had realized that "nothing can save us that is possible," a not-so-original conclusion that has worked itself into the founding story of many religions and philosophical systems and that still leads mortals everywhere to follow the advice of the churches that assist them in their vales of tears.[2]

And yet pain alone, singular pain without the prospect of pleasure would have promoted the avoidance of suffering but not the seeking of well-being. Ultimately, we are puppets of both pain and pleasure, occasionally made free by our creativity.

NATURAL CONSCIOUSNESS

Unannounced and unaccompanied by a proper definition, the word "consciousness" has acquired multiple meanings and become a bit of a linguistic nightmare. This young English word did not even exist in the time of Shakespeare and has no direct counterpart in Romance languages; in French, Italian, Portuguese, and Spanish, one has to make do with the equivalent of "conscience" and use context to clarify which meaning of "conscience" the speaker is after.[1]

Some of the varied meanings of consciousness relate to the optics of the observer/user. Philosophers, psychologists, biologists, or sociologists look at consciousness distinctly. So do ordinary people who hear, night and day, that certain problems are or fail to be "in their consciousness" and who must wonder if consciousness is the erudite label for being awake or attentive or simply having a mind. Yet quietly, hiding under its cultural baggage, there is an *essential meaning* of the word "consciousness," one that contemporary neuroscientists, biologists, psychologists, or philosophers can recognize, even though they approach the phenomenon with varied methods and explain it in different ways. For all of them, more often than not, "consciousness" is a synonym of *mental experience.* And what is a mental experience? It is a state of *mind* imbued with two striking and related features: the mental contents it displays are *felt,* and those mental contents adopt one singular *perspective.* Further analysis reveals that the singular perspective is that of the particular organism within which the mind inheres. Readers who detect a kinship between the notions of "organism perspective," "self," and "subject" will not be wrong. Nor will they be wrong when they realize that "self," "subject," and "organism perspective" correspond to something quite tangible: the reality of "ownership." The "organism owns its particular mind"; the mind belongs to its particular organism. We—me, you, whoever is the conscious entity—own an organism animated by a conscious mind.

To make these considerations as transparent as possible, we need to be clear about the meaning of a few terms: *mind, perspective,* and *feeling. Mind,* as defined earlier, is one way of referring to

the active production and display of images arising from actual perception or from memory recall
or from both. The images that constitute a mind
flow in a never-ending cortege and, as they do so,
describe all sorts of actors and objects, all sorts
of actions and relationships, all sorts of qualities
without and with symbolic translations. Images, of
every kind—visual, auditory, tactile, verbal, and so
forth—individually or in combination, are natural
vehicles of knowledge, they *transport knowledge,*
they explicitly signify knowledge.

Perspective refers to "point of view," provided
there is no doubt that when I use the word view I
do not mean vision only. The consciousness of blind
people also has a perspective, but it has nothing
to do with seeing. By point of view I mean something more general: the relation *I* hold not just to
what *I* see but also to what I hear or touch and,
importantly, even to what *I* perceive in my own
body. The perspective I am talking about is that of
the "owner" of the conscious mind. It corresponds,
in other words, to the perspective held by a living organism *as expressed by the images that flow
within its own mind* when it operates inside that
same organism.

But we can go a bit further in our search for
the origin of perspective. Relative to the world

around us, the standard perspective of most living organisms is largely defined from *the head* of those organisms. In part this is due to the placement of sensory probes—of sight, sound, smell, taste, and even balance—at the top (or front end) of the body. And of course we, sophisticated creatures, also know that the brain is in the head!

Curiously relative to the world inside our organisms, perspective is provided by feelings that unequivocally reveal the natural link between mind and body. Feelings let the mind know, automatically, without any questions being asked, that mind and body are together, each belonging to the other. *The classic void that has separated physical bodies from mental phenomena is naturally bridged thanks to feelings.*

What else do we need to say about feeling in the context of consciousness? We need to assert that self-reference is not an optional feature of feeling but a defining, indispensable one. And we can venture further: we can declare feeling a foundational component of standard consciousness.

In case we get distracted by the saga of the significance of feelings, we also need to recall that all feelings are devoted to mirroring the state of life within a body, whether that state is spontaneous or has been modified by an emotion. This applies fully to all feelings that participate in the process of generating consciousness.

In conclusion, the feelings that are continuously displayed in a mind and are so integral to the making of consciousness have two sources. One source is the never-ending business of running life within the body, which inevitably reflects its ups and downs—well-being, malaise, hunger for food and air, thirst, pain, desire, pleasure. As we saw earlier, these are examples of "homeostatic feelings." The other source of feelings is the collection of emotive reactions, weak or strong, that mental contents frequently prompt—the fears, joys, and irritations that visit us any day. Their mental expressions are known as "emotional feelings," and they are part of the multimedia production that constitutes internal narratives. The feelings endlessly generated by these two mechanisms also become incorporated in the narratives, but they are, to begin with, devices in the generation of the conscious process. In fact the homeostatic variety of feelings helps build the ground zero of our beings.[2]

Consciousness, then, is a *particular state of mind* resulting from a biological process toward which multiple mental events make a contribution. The operations of the body's interior signaled via the interoceptive nervous system contribute the *feeling component,* while other operations within the central nervous system contribute imagery describing the world around the organism as well as its musculoskeletal frame. These contributions converge, in a regimented way, to produce something quite complex and yet perfectly natural: the encompassing mental experience of *a living organism caught, moment after moment, in the act of apprehending the world within itself and, wonder of wonders, the world around itself.* The conscious process takes life within an organism, as expressed in mental terms, and locates it within its own physical boundaries. Mind and body are given joint property of this ensemble, complete with notarized title, and they relentlessly celebrate their luck, good or bad, until they fall asleep.

THE PROBLEM OF CONSCIOUSNESS

Different branches of psychology—aided by general biology, neurobiology, neuropsychology, cognitive science, and linguistics—have made extraordinary progress in the elucidation of perception, learning and memory, attention, reasoning, and language. They have also made significant progress in the understanding of the affects—drives, motivations, emotions, feelings—and of social behaviors.

There is nothing transparent about the biological structures or the processes behind any of these functions, whether they are approached from their public manifestations or from a subjective perspective. It has taken hard work, invention, and a convergence of theoretical efforts and laboratory methods to advance the science of these varied problems. It is thus surprising to realize that consciousness has been discussed as if it stood apart and had been accorded special status, a unique problem, not just difficult to approach, but unsolvable. Some authors on consciousness have sought to overcome the impasse by advancing an extreme proposal, known as "panpsychism." Panpsychists speak about consciousness and mind as if they were interchangeable, something quite problematic. Even more problematic is the fact that they see mind and consciousness as ubiquitous phenomena, present in all living things, as part and parcel of the life state. All single-cell organisms and all plants would be contemplated by their share of consciousness. And why stop at living things? For some, even the universe and all the stones in it are regarded as conscious and minded.[1]

The reasons why these proposals were advanced have to do with an unjustified position, namely that what worked to understand other aspects of mind was insufficient to solve the problem of consciousness. I see no evidence that such is the case. General biology, neurobiology, psychology, and philosophy of mind contain the tools necessary to solve the problem of consciousness and even go a long way toward solving the deeper and underlying problem of the fabric of mind itself. And physics can step in to help as well.

A major issue in consciousness studies concerns what is now commonly known as the "hard problem," the designation that the philosopher David

Chalmers introduced in the literature.[2] An important aspect of the problem refers, in his own words, to "Why and how do physical processes in the brain give rise to conscious experience?"

In brief the problem concerns the alleged impossibility of explaining how a physicochemical device known as the brain—made of *physical objects* known as neurons (billions of them) interconnected by synapses (trillions of them)—could generate *mental states,* let alone *conscious* mental states. How could the brain generate mental states unfailingly connected to a specific individual? And how could those brain-generated states *feel like something,* as the philosopher Thomas Nagel believes they should?[3]

The biological formulation of the hard problem, however, is unsound. Asking why should physical processes "in the brain" give rise to conscious experience is the wrong question. While the brain is an indispensable part of the generation of consciousness, nothing suggests that the brain generates consciousness alone. On the contrary, the non-neural tissues of the organism's body proper contribute importantly to the creation of any conscious moment and must be a part of the problem's solution. This happens most notably via the hybrid process of feeling, which we regard as a critical contributor to the making of conscious minds.[4]

What does it mean to say "I am conscious"? At the simplest level imaginable, it means to say that my mind, at the particular moment in which I describe myself as conscious, is in possession of knowledge that spontaneously identifies me as its proprietor. Foundationally, the knowledge concerns *myself* in varied ways: (a) my body, about which I am continuously informed in greater or lesser detail via feeling, (b) along with facts that I recall from memory and that may pertain (or not) to the perceptual moment and are also part and parcel of myself. The scale of the knowledge fest that renders minds conscious varies depending on how many honorable guests attend, but certain guests are not only honored but obligatory. Let me identify them: first, *some knowledge about the current operations of my body;* second, *some knowledge as retrieved from memory, about who I am at the moment and about who I have been, recently and in the long ago past.*

I will not fall in the trap of saying consciousness is this simple, because it is not simple at all. There is nothing gained by underestimating the complexity generated by so many moving parts and articulation points. As complicated as consciousness is, however, it does not appear to be—or have to remain—mysterious or impossible to figure out in terms of what it is made of, mentally speaking.

I am full of admiration for how our living organisms—the parts that we call neural and those that we tend to ignore and dismiss as "the rest of the body"—have concocted the processes that result in mental states imbued with feeling and a sense of personal reference. But admiration does not require the invocation of mystery. The notion of mystery and the idea that a biological explanation lies beyond us do not apply. Questions can find answers, and puzzles get resolved. Still, one is filled with awe at what the combination of several relatively transparent functional arrangements has ended up producing for our benefit.[5]

WHAT IS CONSCIOUSNESS FOR?

This is an important question, but few people ask it seriously. The idea that consciousness would be useless has been floated, but if consciousness would serve no purpose, would it still be around? In general, useful functions are maintained and honed in biological evolution, while useless ones tend to be discarded, that being the job of natural selection. To be sure, consciousness is not useless.

First, consciousness helps organisms govern their lives in keeping with the strict requirements of life regulation. This is true of many nonhuman species that preceded us and dramatically true of humans. This should not be surprising. After all, one of the foundations of consciousness is feeling, whose purpose it is to assist with the governance of life in line with homeostatic requirements. One might say, in an effort to give the birth of consciousness its due, that there is a chronology, that feeling emerged in evolution just one half step ahead of consciousness, that feeling is, literally speaking, a stepping-stone for consciousness. The reality, however, is that the functional value of feelings is tied to the fact that they are unequivocally referred to their owner organism and inhabit their owner-organism's mind. Feelings gave birth to consciousness and gifted it generously to the rest of the mind.

Second, when organisms are very complex—certainly by the time they have nervous systems capable of supporting minds—consciousness becomes an indispensable asset *in the struggle to govern life successfully.*

It is possible for independent living organisms to proceed successfully, without minds or consciousness, as we see in bacteria and plants. Their problems of existence and persistence can be solved with far less panache by a powerful *non-minded competence,* a sort of sneaky and very intelligent forerunner of mind and consciousness combined. I call this non-conscious competence "sneaky" because it ends up governing the life of non-conscious creatures quite well, without the athletic trappings of subjective experiences.

But we must note that, importantly, while conscious minds generate explicitly intelligent governance, they are also helped by non-explicit intelligence, as needed. Life is not possible when it runs unattended and ungoverned. It needs to be managed. Either a conscious mind or a non-explicit competence is indispensable for good life governance, but the full scope of intelligent management, from non-conscious to conscious, is not required by all species.

Because consciousness connects the mind indelibly to a specific organism, it assists the mind in making a pressing case for the particular needs of that organism. And when organisms can mentally describe the degree of their needs and can apply knowledge to respond to those needs, then theirs is the universe to conquer. Conscious minds help organisms clearly identify what is required for their survival, and feel their way through the requirements. Often, depending on the degree of feeling involved, consciousness may demand and even impose a response to the identified needs. Explicit knowledge and reason provide resources not available to implicit forms of competence, which are governed by concealed varieties of intelligence and responsive only to basic homeostasis. Knowledge and creative reasoning invent novel responses to specific needs.

Organisms endowed with conscious minds gain remarkable advantages. In keeping with their degree of intellect and creativity, their field of action widens. They can struggle for life in more varied settings. They can face a larger variety of hurdles and have a better chance of overcoming them. Consciousness expands their habitat.

Organisms with large mental capacities use consciousness—that is, the ownership reference of those mental capacities to their bodies—in their calculations and creative endeavors. Their entire program of behavior benefits from consciousness. Rather than asking why our creative processes should be accompanied by consciousness, we should wonder how any of our best behaviors would be possible—let alone useful—in the absence of consciousness.

MIND AND CONSCIOUSNESS
ARE NOT SYNONYMOUS

It took me a while to realize that part of the problems we face when we debate consciousness comes from a serious confusion. Consciousness is a distinctive state of mind, but the words "consciousness" and "mind" are often used as if they were synonymous and corresponded to the same process. Pressed hard on the point, the "misusers-confusers" may admit as much, but they let the critical distinction fall by the wayside. They and their listeners become unable to envision the central mechanism of consciousness as a *modification* of the primary process of mind.

The confusion is a consequence of the "composition problem." The constitutive components of complex phenomena are difficult to glean under the functional envelope that obscures them. Referring to "conscious minds" instead of "consciousness"—as I do in the subtitle of this book—is helpful because "conscious" qualifies "minds" and serves notice not all mind states are necessarily conscious, that there are *components* involved in the making of consciousness.

In my proposal consciousness is an *enriched* state of mind. The enrichment consists in *inserting additional elements of mind within the ongoing mind process*. These additional mind elements are largely cut from the same cloth as the rest of the mind—they are imagetic—but thanks to their contents they announce firmly that *all the mental contents to which I currently have access belong to me, are my thing, are actually unfolding within my organism*. The addition is *revelatory*.

Revealing mental ownership is first and foremost accomplished by feeling. When I experience the mental event we call pain, I can actually localize it to *some part of my body*. In reality, the feeling occurs in *both* my mind and my body, and for a good reason. I own both, they are located within the same physiological space, and they can interact with each other.

The manifest ownership of mental contents by the integrated organism where they arise is the distinctive trait of a *conscious* mind. When this trait is absent or not dominant, the simpler term *mind* is the appropriate descriptor.

The mechanisms involved in enriching a mind with a firm connection to its rightful owner organism consist in inserting in the organism's mental flow the contents that connect *mind* and *organism owner* unequivocally. They occur at the level of systems. They should not be regarded as a mystery.

My solution to the problem of consciousness does not imply that all the biological mechanisms behind consciousness are clarified. Nor does it imply that states of consciousness are all equivalent in scope and grade. There is a distinction to be made between my conscious mind when I wake up from deep sleep—and all I barely know is who I am and where I am—and the conscious mind that helps me think for hours through a complicated scientific problem. But my solution to the consciousness problem is applicable and decisive in both cases. For a conscious mind to emerge, I need to enrich a plain mind process with knowledge that pertains to my organism and that identifies me as the owner of my life, my body, and my thoughts.

Both the simple, conscious mind process focused on a mundane problem and the rich, panoramic, conscious mind process that encompasses a vast amount of history depend on an initiation rite: *the identification of an "owner-mind" which requires the placement of that mind in the setting of its body.*

BEING CONSCIOUS IS NOT THE SAME AS BEING AWAKE

Being conscious and being awake are often regarded as the same, and yet consciousness and wakefulness are quite distinct. To be sure, consciousness and wakefulness are related. We know that when organisms fall asleep, their consciousness is usually turned off, although we must also remember a blatant exception to that rule: when we are sound asleep, consciousness returns during our dreams, creating a rather bizarre situation. We are asleep *and* we are conscious. Moreover, in some variations of the state of coma, patients are apparently unconscious, and yet their electroencephalogram suggests that they remain technically awake. I know that this sounds complicated and confusing, but I can attest that once the fog of these variations lifts, we can confidently say that consciousness is not mere wakefulness.[1]

We should think of wakefulness as the operation

that allows us to "inspect" images, a sort of turning on the lights on the set. But the wakefulness process is not involved in putting together the procession of images in our minds, nor is it concerned with telling us that the images we are inspecting are ours.

As we discovered earlier in the discussion on minds, the ability to "sense" or "detect"—a touch, a rise in temperature, a vibration—is not to be confused with mind or consciousness either.

CONSCIOUSNESS (DE)CONSTRUCTED

Why do I believe there is a plausible solution for the problem of consciousness? First, because I can envision a means whereby mental contents are transparently connected to a feeling subject and the feeling subject assumes ownership of those contents. Second, because the means I envision calls for the use of a physiological mechanism whose status, at the level of systems, is reasonably understood.

Consciousness is constructed by adding to the flow of mental images we call mind an extra set of mental images that express *felt* and *factual* references to the mind's owner. Mental images, both conventional and hybrid, such as feelings, carry and convey meanings that are the key ingredients of consciousness, just as they are the key ingredients of plain minds. No previously unknown phenomenon is invoked or required and no mysterious stuff needs to be added to the brew of images in order to render the ensemble conscious. The key to consciousness resides in the *contents* of the enabling images. It resides in the *knowledge* those contents naturally provide. All the images need to be is informative so that they can help identify their owner.

Proposing a solution for consciousness that does not appeal to the unknown and mysterious does not mean that the solution is "simple"—it is not—and does not imply that all problems related to the operation of conscious minds are solved; they are not. What happens in our organisms when we are experiencing a performance of Wagner's *Ring,* physiologically speaking, is not for the fainthearted, musically, theatrically, and biologically speaking.

The image contents of minds hail largely from three principal universes. One universe concerns *the world around us.* It yields images of the objects and actions and relationships present in the environment that we occupy and that we continuously scrutinize with the external senses—visually and auditorily, by touch and by smell and taste.

The second universe concerns *the old world inside us.* This world is "old" because it contains evolutionarily ancient internal organs in charge of metabolism: viscera such as the heart, lungs, stomach, and guts; large and independent blood vessels as well as those located in the depth of the skin; endocrine glands, sexual organs, and so forth. This is the universe that gives rise to feelings, as we have seen in the sections on affect. The images that are part of feelings also correspond to actual objects, actions, and relationships but with some monumental distinctions. First, the objects and actions are located *within* our organisms, in the visceral interior that sits largely inside the chest, abdomen, and head, as well as in the extensive viscera that inhabit the thick of the skin, throughout the body, traversed by blood vessels with their smooth muscular walls.

Moreover, rather than merely representing the shapes or actions of internal objects, the images from the second universe principally represent *states* of the objects relative to their function within our living economy.

Lastly, the processes in the old world universe shuttle back and forth between the actual "objects," for example, the viscera, and the "images" that represent them. There is a continuous interaction between the sites where the body actually changes and the "perceptual" representation of those changes. This is a thoroughly hybrid process, simultaneously "of the body" and "of the mind"; it allows the images on the mind side to be updated following the alterations occurring in the body and be changed accordingly. Of note, relative to the life process, the images represent qualities and their momentary value or valence. The *state* and *quality* of the actual objects and actions in the interior are the stars. It is not the actual violins or trumpets that steal the show; it is the *sounds* they make. In other words, feelings are not reducible to fixed imagetic patterns; they concern "ranges" of operation.

A third universe of mind also pertains to a world within the organism but involves an entirely different sector: *the bony skeleton, the limbs and the skull, body regions that turn out to be protected and animated by skeletal muscles.* This sector of the interior provides both *frame and support* for the whole organism and anchors the external movements executed by skeletal muscles, including those we use for locomotion. This entire frame serves as reference for everything else that goes on in the first and second universes. Interestingly, from an evolutionary standpoint, this sector of the interior is not as old as the visceral one and does not share the same peculiar physiological traits. There is nothing soft about this "not-so-old interior." Solid bones and tough muscles make good scaffolds and good frames.

EXTENDED CONSCIOUSNESS

The idea that minds can be rendered conscious once feeling is present and the subject identified, may be surprising, at first glance, which is not a problem. However, the idea that the explanation of consciousness that I am offering may be regarded as too "small" for the "importance" of the phenomenon *is* a problem and needs to be addressed.

The problem, as I see it, is actually caused not by the explanation but rather by expectations that have been associated with traditional, vague, and inflated notions of what consciousness is supposed to be, as distinct from what consciousness actually *is* and *does*. Earlier I noted the unique evolutionary role of consciousness and the fact that it has been indispensable in the history of humanity. Moral choice, creativity, and human culture are conceivable only in the light of consciousness. These facts, however, are entirely compatible with the scale at which I place the critical mechanism behind consciousness.

One reason why the explanation I advance may sound modest at first has to do with the notion of *Extended Consciousness,* a concept that I introduced when I first began studying the problem and of which I used to be rather fond.[1] The designation "Extended" applied to what I saw as the large-scale variety of consciousness, the one meant to encompass our experience of reading Marcel Proust and Leo Tolstoy and Thomas Mann and listening to Mahler's Fifth: wide, tall, rich, long, containing multitudes of humanity and their respective habitats, drawing on the past that we have committed to memory, playing creatively with our stores of knowledge, and projecting itself into the possible future.

The problem, as I see it today, is that I should have talked, all along, about Extended *Mind* rather than Extended Consciousness. The fundamental mechanism whereby images are rendered conscious remains the same when the device is applied to a million images or to only one. What does change is the scale and capacity of our mind processes as demanded by the quantity of materials we recall and are working on and by the forces of attention that are called to intervene, and as, bit by bit, entire canvases of music, literature, painting, and cinema are *mentally encompassed* and made to belong to us, that is, *rendered conscious.*

WITH EASE, AND YOU BESIDE

I used to think of Emily Dickinson's famous poem as an ode to consciousness, but now I see it as making penetrating observations on the human mind.[1] Consider the first four lines:

> *The brain is wider than the sky,*
> *For, put them side by side,*
> *The one the other will include*
> *With ease, and you beside.*

Dickinson intuits the need for "you" in the process of making a conscious mind—that being me or any other individual—but her focus is on the *scale* of that mind. How come the visual panorama and the auditory scene that I am currently beholding are so much larger than the modest width of my brain? That is what she wants to know.

The brain had to be wider than the sky—by which she meant larger than the skull—because the brain could contain not just the world around us but *you,* beside. As Dickinson well knew, however, neither the world nor we could actually fit inside the skull. First, we and the world had to be miniaturized, rescaled to brain proportions. Once the rescaling was accomplished, we and our thoughts were allowed to inflate to the size of the near and far universe while still fitting in the head.

Dickinson was candidly committed to an organic view of mind and to a modern conception of the human spirit. And yet, in the end, what turned out to be wider than the sky was not the brain but life itself, the begetter of bodies, brains, minds, feelings, and consciousness. What is more impressive than the entire universe is life, as matter and process, life as inspirer of thinking and creation.

THE REAL WONDER OF FEELINGS

Feelings again, must we? We must indeed. Feelings protect our lives by informing us of dangers and opportunities and by giving us the incentive to act accordingly. Those are natural wonders, no doubt, but feelings offer another wonder, the one without which their guidance and incentives would not be heeded. They provide the mind with facts on the basis of which we know, effortlessly, that whatever else is in mind, at the moment, also belongs to us, is happening in us. Feelings allow us to experience and become conscious, to unify our mental holdings around our singular being. Homeostatic feelings are the first enablers of consciousness.

The critical facts that feelings offer to the mental process concern specifics about the organism's interior continuously modified by homeostatic adjustments. They show that the entire process is occurring in a mind that is part of the organism within which the homeostatic adjustments are occurring! The mind "belongs" to "its" organism.

The feelings that make consciousness possible are not in a class apart. They juxtapose two principal phenomena: (1) images of the interior, which detail the homeostatically driven alterations of the organism's internal configurations; and (2) images that detail the *interactions* between the maps and their body sources and that, by so doing, naturally reveal that the mappings are made inside the organism they represent. The discovery of ownership results from the mutual and transparent influences of the organism state and of the images generated in that organism; ownership is consequent to the patent fact that one process—the fabrication of mental images—occurs inside the other—the organism.

The fact that the organism owns the mind has an intriguing consequence: all that occurs in the mind—the maps of the interior and the maps of the structures, actions, and spatial positions of other organisms/objects that exist and take place in the surrounding exterior—is constructed, of necessity, by adopting *the organism's perspective*.

THE PRIORITY OF THE WORLD WITHIN

When people casually refer to consciousness, they are usually thinking of the external world first. They often equate being conscious with the ability to represent the world of their surroundings. This is understandable because the world external to us is so disproportionately favored in our minds. But why is that so? Because mapping the world around us is essential to govern our interactions with that world in ways that can be favorable to our lives. Still, while this process helps reveal what can be known and used to our advantage, it does not suggest, let alone explain, how or why we are conscious of the material we have mapped in images, in other words, why we know that we know. *To be knowledgeable* and *conscious, we need to "connect" or "refer" objects and processes to our own organisms, to ourselves. We need to establish our organisms as surveyors of the objects and processes.*

We become conscious of our existence and of

our perceptions when we use knowledge to establish reference and ownership.

We only come to know that we know—which really means that we only come to know that *each of us, individually,* is in the possession of knowledge—because we are simultaneously informed about two other aspects of reality. One aspect concerns the states of our ancient chemical and visceral interior, expressed in the hybrid process called feeling. Another aspect is the spatial reference provided by our musculoskeletal interior, especially the stable frame that anchors the edifice of our selfhood.

A GATHERING OF KNOWLEDGE

One might try to see the process of constructing "consciousness" as that of a successful building contractor who gathers the materials and the artisans needed for his project. Consciousness gathers together the bits of sapience that reveal, by dint of their coincident presence, the mystery of belonging. They tell me—or you—sometimes in the subtle language of feeling, sometimes in ordinary images or even in words translated for the occasion, that yes, lo and behold, it is me—or you—thinking these things, seeing these sights, hearing these sounds, and feeling these feelings. The "me" and "you" are identified by mental components and body components. It makes no difference provided that the connection between mental events and overall body physiology has been robustly established. The world can come to you, says your contractor in charge of consciousness, because your living organism—your whole organism, not just your brain—is an open stage where a relentless play is being played out, for your benefit. The materials for the construction, brick after brick, are just knowledge and not different from those in the rest of mind. Its substrate is images and more images, including those hybrid images that rely on brain-body interactions and come complete with tugs and pullings: the "images" we call feelings. The bits of knowledge that are piled on top of the running mental tracks, those castellations of images that help describe the moment of our lives, our living time, those bits of knowledge are a relentless demonstration of being.

Consciousness is a gathering of knowledge sufficient to generate, in the midst of flowing images, automatically, the notion that the images are *mine*, are happening in *my* living organism, and that the mind is . . . well, *mine* too! The secret of consciousness is gathering knowledge and exhibiting that knowledge as a certificate of identity for the mind. Consciousness is not a mere integration of mental elements, although integration does have a role to play when consciousness is conferred upon large numbers of images.

In retrospect, an error that has been repeatedly committed in the quest for consciousness has been to treat it as a "special" function, even a separate "substance," a fragrance wafting over the mind process but unconnected to it or to its underpinnings. Even those of us who imagined less outrageous solutions to the problem made it more mysterious than it needed to be.[1]

INTEGRATION IS NOT
THE SOURCE OF CONSCIOUSNESS

When we describe ourselves as conscious of a particular scene, we require a considerable integration of the components of the scene. There is no reason to expect, however, that integration alone, no matter how abundant, would be responsible for consciousness. Increased integration of mental contents, over larger amounts of flowing imagetic material, delivers a larger scope of conscious material, but I doubt that consciousness is explainable by the "tying together" of the contributing contents. Consciousness does not spring forth just because mental contents are appropriately assembled. I would suggest that the result of integration is an enlargement of the mental scope. What does begin to engender consciousness is the enrichment of the mental flow with the sort of knowledge that points to the organism as the proprietor of the mind. What begins to make my mental contents conscious is identifying

ME as owner of the current mental holdings. Ownership knowledge can be obtained from specific facts and, quite directly, from homeostatic feelings. Easily, naturally, and instantaneously, as often as needed, homeostatic feelings *identify* my mind with my body, unequivocally, no extra reasoning or calculation needed.[1]

CONSCIOUSNESS AND ATTENTION

Consciousness is not unlike milk and eggs. It comes in grades that largely correspond to the kind and amount of mental material made conscious at any time. The grading, however, is complicated by a curious interplay between the kind of material present in mind and the attention one dedicates to it. For example, as I started writing this page, I was quite focused on the ideas I wanted to convey. But something happened as I pondered matters; I also pressed the remote control of the CD player and on came the sound of a disc I had selected earlier in the day. The scope of my conscious mind enlarged considerably to accommodate the new material, but I was now divided between the topic of my writing—the scope of consciousness!—and a demanding comparison between the way the particular pianist I was hearing resolved certain phrases and how another and older pianist played exactly the same passages. This text demonstrates the consequences:

the primary purpose of my project recessed into the background, still in "conscious mind" but not up front and close, while the music soldiered on to prominence. Not long after, the position of the contents reversed, and I was again writing about consciousness.

I had been distracted but now returned to the proper focus.

It is not reasonable to analyze my distraction in terms of consciousness only or attention only. Both have a say in the matter. The secondary process of enhancing the quality of certain images or their filmic "editing"—how large are the shots selected or how long they take—is technically speaking an issue in the domain of attention. It is also not reasonable, however, to overlook the role of affect in the allocation of "attention" to the materials available for selection into my mind flow. Deciding on how Leif Ove Andsnes differed from Martha Argerich and whereabouts in the piece was suddenly more rewarding—enjoyable—than clarifying my ideas on the scope of consciousness. I allowed that pleasurable task to dominate the proceedings.

None of what transpired above should alter our interpretation of the biological reality: the contents selected for my mind were identified as belonging to me thanks to the foundational feeling process that declared me their sole owner, and thanks to fringe facts that described me in my current position, at my desk, with the sounds booming around me, and the sun setting over the Getty Museum, out there to my right, a bit west and a bit north.

Attention helps manage the abundant production of images in mind. It does so on the basis of (a) the intrinsic physical characteristics of the images, for example, colors, sounds, shapes, relationships; (b) the significance of the images both personally (as established with the help of individual memory) and historically. A mixture of emotive and cognitive responses subsequently governs the time and scale allocated to the images that get to be incorporated in the conscious mental flow.[1]

THE SUBSTRATE COUNTS

One bizarre consequence of the extraordinary success of the computational sciences is the idea that minds, including the human variety, would not depend on the substrate that supports them. Let me explain the idea. I am writing these sentences with a Paper Mate pencil No. 2, on a yellow paper pad, but I might just as well have typed them on an old Olivetti typewriter or on my iPad, or on a laptop. My words would be the same, so would the syntax and punctuation. The ideas and their linguistic interpretation would be independent of the substrate used for communicating them. This may appear reasonable at first glance, but it does not fit the reality of feeling/conscious minds. Can we say that the contents of our minds are independent of the organic substrate that carries them, namely the brain and the living organism of which it is a part? Not really. The narratives we construct, the characters and the events in the narratives, the considerations we make regarding the characters that play in these events, the emotions we attribute to those characters, and those we experience as we watch events unfold and react to them are not independent of their organic substrate. The idea that the contents of our minds stand, relative to the nervous system and to the living organism, in the same way that the text that I am writing stands relative to its many possible substrates—pencil, typewriter, computer—is flawed.

A good part of our mental experience—sometimes most of it—is not strictly confined to the objects, characters, and pratfalls in the narratives that flow forward in our mental stream. A good part also includes the experience of the organism itself, which depends on the state of life in that organism, well or not so well. In the end, our mental experiences are best described as experiences of "being" while "other mind contents" flow along. The "other mind contents" flow in parallel to the "contents of being." Moreover, "being" and "other mind contents" are engaged in a dialogue. One or the other dominates the mental moment depending on how rich the respective descriptions are. The "being" component is permanently present, even when it is not dominant, constructed from non-neural and neural elements. To say that our conscious minds would be substrate-independent would be to say that the edifice of "being" could be dispensed with and that only the "other mind contents" would count. It would be to deny that the foundation of mental experiences is, to begin, the *experience/consciousness* of a *particular kind of organism, in a particular state.*

The substrate counts, it has to count, because *that substrate is the organism of the person who is experiencing the story and reacting to it affectively.* That is also the person whose affective system is being "borrowed" to give some semblance of life to the emotions of the characters being depicted in the story.

LOSS OF CONSCIOUSNESS

The distinguished philosopher John Searle was fond of beginning his lectures on consciousness with a lapidary definition that signified his satisfactory resolution of the problem. There is no mystery to consciousness, he would say. Consciousness is merely what disappears when you go under anesthesia or when you reach deep, dreamless sleep.[1] This is an attractive way of beginning a lecture, for certain, but it does not satisfy as a definition of consciousness, and it is misleading in relation to anesthesia.

True enough, consciousness is not available in dreamless sleep or during anesthesia. Consciousness is nowhere to be found in a state of coma or in a persistent vegetative state, it can be compromised under the influence of a variety of drugs and alcohol, and it slips from us momentarily when we faint. Consciousness is *not* lost, although it may appear to be so, in a devastating condition known as locked-in syndrome in which neurological patients are unable to communicate and *seem* unaware of self and surroundings but are in reality perfectly conscious.

Unfortunately, neither anesthesia nor the neurological conditions that impede consciousness achieve that result by specifically targeting the mechanisms for constructing a conscious mind that I have been describing. Anesthesia and pathological states are rather blunt tools.[2] They *target functions on which normal consciousness depends* rather than consciousness itself. As I indicated earlier, the serious anesthetics used in surgery are fast instruments that instantly suspend *sensing/detecting*, the interesting function to which I called attention when we discussed *unminded and non-conscious bacteria*. The evidence in support of this statement is clear. Bacteria are able to sense and so are plants, but neither are minded or conscious. Nonetheless, anesthetics suspend their sensing and place them in a literal hibernation while obviously doing nothing *specifically* against consciousness, a function neither bacteria nor plants had, to begin with.

Sensing does not entitle us to minds or consciousness; but in the absence of sensing we cannot build up the operations that gradually enable plain minds, feelings, and self-reference, the ingredients that eventually permit *conscious minds*. In brief, as I see it, anesthetics do not alter consciousness primarily; they alter sensing. That they ultimately preclude the ability of putting conscious minds together is a very useful and practical effect because we are interested in having surgical procedures without ever being conscious of pain.

Alcohol, plenty of painkillers, and numerous drugs that humans have used for millennia for all sorts of personal and social reasons provide another example of *interference* with the normal process of assembling a conscious mind, and they are a bit closer to the mark. They can jitter the final assembly of consciousness or preclude a critical step. The connection is a curious one. The long-standing personal and social reasons that explain the use and abuse of substances such as narcotics and alcohol are tied to their effects on the physiology of feeling. The users are not interested in modifying consciousness, especially, but rather in modifying certain homeostatic feelings such as pain and malaise—which we all want to see banished from our beings—and well-being and pleasure, which we all wish to maximize, and then some, if at all possible.

Clearly, any drug capable of penetrating the den of homeostatic feelings has found a way into the machinery of consciousness, which is grounded, in no small measure, on the homeostatic feeling process. This is a connection that explains the interference of drugs in the process of consciousness.

And what about syncope, otherwise known as fainting? We faint because blood flow to the brain stem and cerebral cortex suddenly drops below a prohibitive level. A large swath of brain operations is suspended as a consequence of insufficient oxygen and nutrients being delivered to neurons in brain regions that contribute importantly to the assembly of feelings, especially in the brain stem. Information from the organism's interior is suddenly kept out of the central nervous system, and the contribution of feelings to consciousness is rudely interrupted. Muscular tone is as much compromised as the sense of self and surroundings, and that is why the victim swoons and sways and falls to the ground, just as some notable patients did during the magisterial demonstrations of Jean-Martin Charcot at the Salpêtrière Hospital in Paris. Charcot was one of the pioneers of neurology and psychiatry during the second half of the nineteenth century. He became famous for studying a disease that no longer exists: hysteria. Sigmund Freud attended some of his lectures, to great profit.

Connecting the loss of consciousness to the brain stem is a modern view, advanced by another historical figure, the neurologist Fred Plum.[3] My interpretation of why the brain stem is a key to consciousness connects with the notion that feelings are expressions of homeostatic operations and that they are essential to producing consciousness. We know today that important components of the machinery behind both homeostasis *and* feelings are housed in the upper sector of the brain stem, above the level of the trigeminal nerve entry and, quite specifically, in the back portion of that sector (the area marked as B in figure IV.1. Of interest, damage to this brain stem sector is a well-established cause of coma.[4] Curiously, damage to the front portion of this same sector (marked as A in the same figure) does *not* cause coma, does *not* compromise consciousness at all, and produces instead the condition of "locked-in," to which I referred earlier. The tragic victims of this syndrome are awake and alert and conscious but largely unable to move and thus drastically reduced in their ability to communicate.

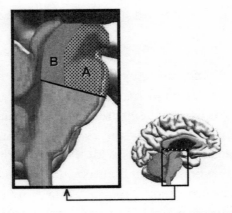

Figure IV.1: Detail shows an enlargement of the brain stem region. Damage within the sector marked B is consistently associated with loss of consciousness. Damage within sector A is associated with motor impairments.

THE CEREBRAL CORTICES
AND THE BRAIN STEM IN
THE MAKING OF CONSCIOUSNESS

It has been said that the posterior sensory cortices, unlike the anterior, prefrontal ones, are the natural basis of consciousness. There is a whiff of truth in that idea, but only a whiff. Reality is more complicated.

The posterior sensory cortices—largely located in the back part of the brain—include the so-called "early" sensory cortices of vision, hearing, and touch; they are lead fabricators and exhibitors of visual, sound, and tactile images. But the so-called "higher order" association cortices of each sensory modality, which intersect at the junction of the temporal and parietal lobes (TPJ) are also involved in image-making and in the assembly of composite images (see figure IV.2 where the main cerebral cortices are identified).

In effect, the entire lateral and posterior cor-

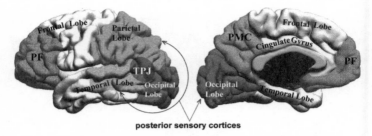

posterior sensory cortices

Figure IV.2: The principal regions of the human cerebral cortex. PF = Prefrontal Cortex; PMC = Postero Medial Cortices; TPJ = Temporal Parietal Junction.

tical territory is involved in image-making and image-display, which is the same as saying that it is involved in making minds. But what about consciousness we need to ask? Does this brain territory also contribute to making the respective minds conscious? In part, at least, that seems to be the case. Consciousness being an image-based process, it requires plenty of images as substrate, something the posterior sensory cortices provide abundantly. Some regions of these cortices help with the integration of images and probably orchestrate their sequencing as they become conscious. But what makes us conscious of the images that the posterior cortices fabricate and sequence with ease is the *addition of knowledge certifying the ownership of those images,* the discovery that those images belong to a particular organism with unique physical traits and a unique mental history anchored in memory. For those who expect the posterior sensory cortices to be the sole providers of consciousness, this is where the trouble begins: *the primary mechanism for conferring ownership upon images is the presence of homeostatic feelings, but this presence does not depend on the posterior cortices primarily.* As we have seen, feelings are hybrid processes whose images depict back-and-forth interactions of the interoceptive nervous system with the actual viscera in our interior.

The structures responsible for feelings are located in (1) the peripheral component of the interoceptive system, (2) the brain stem nuclei, (3) the cingulate cortex, and (4) the insular cortices. The inputs and overall design of the insular region allow it to integrate representations of multiple sources of interior processes, including those that correspond to interactions of sensors with actual viscera. The higher levels of the feeling process probably depend on the insular cortex region, a sector that completes and refines the job accomplished by numerous prior structures in a long chain that begins in the spinal ganglia and the spinal cord and that continues in the brain stem, notably in the parabrachial nucleus, the periaqueductal gray, and the nucleus of the trac-

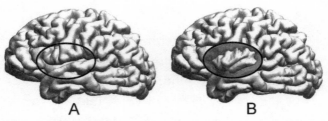

Figure IV.3: The insular cortex is buried in the depth of each hemisphere. The oval mark in panel A marks the cortical territory under which the actual insular cortex is located, as shown in panel B.

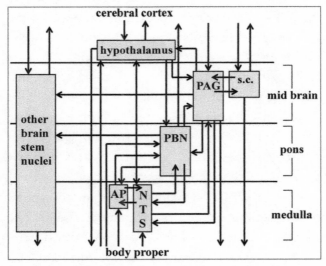

Figure IV.4: Diagram of the main brain stem structures involved in affective processes, their interconnections, sources of input, and targets of output. PAG = periaqueductal gray; s.c. = superior colliculi; PBN = parabrachial nucleus; AP = area postrema; NTS = nucleus tractus solitarius.

tus solitarius. Together, the insular cortex and the subcortical components that feed into it constitute an "affect complex" (see figures IV.3 and 4).

The critical question, at this point, is how do these two sets of structures—the posterior sensory cortices and the "affect complex"—combine to produce conscious minds? I envision two possibilities. One calls for actual neural projections from the "affect complex" to the "posterior sensory set" and vice versa. The other possibility calls for approximate simultaneity of activations in the two sets, resulting in the production of a time-based ensemble. In either option, the ultimate realization of a conscious mind depends on *both* sets of brain structures; we cannot "localize" consciousness to one or the other set. Moreover, one other sector of the cerebral cortices appears to play a role in coordinating the conscious mind processes. The sector is known as the PMC (the Postero Medial Cortices; see figure IV.2). It encompasses cortices largely located in medial (internal) and posterior surfaces of the cerebral hemispheres. This region may possibly direct the participation of other cerebral cortices in the making of a conscious mind.

And what about the frontal cortices? Are they involved in the making of consciousness? The answer is that the anterior frontal or prefrontal cortices (PF in figure IV.2) do *not* have a primary role in producing conscious minds. Classic brain lesion studies in humans have shown that the damage or even the surgical ablation of the prefrontal cortices do not compromise the basic process of making minds conscious. The anterior frontal cortices are involved in image manipulation and promote the activation, sequencing, and spatial positioning of images fabricated in the posterior sensory cortices, the orchestrating role that some regions of the posterior sensory cortices and of the PMC also play. The frontal cortices appear to be instrumental in assembling the vast mental panoramas that the process of consciousness literally illuminates and identifies as ours.

While the frontal sector contributes significantly to intelligent mental operations—reasoning, decision-making, creative constructions—it does not appear to contribute to the essential knowledge enrichment on which basic consciousness depends. It does not authenticate the proprietor of mind, and it does not grant it ownership, but it is instrumental in generating the large-scope *extended mind* that represents human capacities at their peak.[1]

FEELING MACHINES AND CONSCIOUS MACHINES

Robotics are the ultimate expression of artificial intelligence (AI), and let me begin by saying that the label "artificial" could not be more appropriate. There is nothing "natural" about the intelligence of the devices that make our lives so efficient and comfortable, and there is nothing "natural" about the construction of those devices. Still, the brilliant inventors and engineers who made AI and robotics possible were inspired by natural, living organisms, especially by the smarts with which living creatures solve the problems they face and by the efficacy and economy of their movements.

One might have expected the pioneers of AI and robotics to have sought inspiration from the whole-ness of beings such as we are—full of efficiency and dispatch but also full of feelings about everything that we are efficient and dispatched about, in brief, joyful and even ecstatic about what we do (and are done to) but also frustrated and sad and even pained when the occasion calls for it.

The brilliant pioneers, however, took an economical approach and cut to the chase. They tried to emulate what they regarded as most essential and useful—let's call it plain intelligence—and left out what they probably regarded as superfluous and even inconvenient: *the feeling stuff.* Quite possibly they regarded affect as not just quaint but outmoded, something left behind in the otherwise triumphant march toward clarity of thought, exact problem solving, and precise action.

In the light of history, their choice is understandable. It has unquestionably produced many excellent results and wealth to match. My qualification, however, is that in proceeding the way they did, the pioneers revealed a significant misconception regarding human evolution and, by so doing, limited the scope of AI and the respective robotics in terms of their creative potential and ultimate level of intelligence.

The evolutionary misconception should be obvious in light of what we have been discussing in this book. The universe of affect—the feeling experiences derived from drives, motivations, homeostatic adjustments, and emotions—was a *prior histori-*

cal manifestation of intelligence, highly adaptive and efficient, and was a key to the appearance and growth of creativity. It was several notches up from the hidden and blind competences of bacteria, for example, but shy of the full-fledged human intelligence. Indeed, the universe of affect was the stepping-stone for the higher intelligence that conscious minds gradually developed and expanded. The universe of affect was a source and an instrument in the development of the gradual autonomy we humans conquered.

It is time to recognize these facts and time to open a new chapter in the history of AI and robotics. It is apparent that we can develop machines that operate along the lines of "homeostatic feelings." What we need, in order to do so, is to provide robots with a "body" that requires regulations and adjustments in order to persist. In other words, we need to add, almost paradoxically, a degree of vulnerability to the robustness that is so prized in robotics. Today this can be achieved by placing sensors throughout the robot's structure, having them detect and register the more or less efficient states of the body, and integrating the corresponding information. The novel technologies of "soft robotics" enable this development by trading rigid structures for flexible and adjustable ones. We also need to transfer this "sensing and sensed" body influence to the organism components that process and respond to the conditions surrounding the machine so that the most effective—intelligent—response can be selected. In other words, what the machine "feels" in its body will have a say on the matter of responding to the conditions surrounding it. That "say" is meant to improve the *quality and efficiency of the response,* therefore making the robot's behavior more intelligent than it would otherwise be in the absence of guidance from its internal conditions. Feeling machines are not aloof and predictable robots. To some extent, they care for themselves and outsmart their conditions.

Do such "feeling" machines become "conscious" machines? Well, not so fast. They do develop functional elements related to consciousness, feeling being part of the path to consciousness, but their "feelings" are not equal to the feelings of living creatures. The eventual "degree" of consciousness of such machines will depend on the complexity of the internal representations of both the "interior of the machine" and its "surround."

In the appropriate setting, a new generation of "feeling machines" can probably become efficacious assistants to really feeling humans, as hybrids of natural and artificial creatures. No less important, this new generation of machines would constitute a unique laboratory for the investigation of human behavior and mind in a variety of actual realistic settings.[1]

V
*In All Fairness:
An Epilogue*

Life and natural selection are responsible for the multitude of organisms we find around us and for our presence as well. Over billions of years diverse organisms held on to life, through thick and thin, for more or less limited periods of time, and, once their existence reached a natural or accidental terminus, made way and room for other living organisms. Humans, the latecomers in this saga, rather than merely and modestly enduring and prevailing, have become ever more elaborate in their behaviors, created environments to match, and dominated the planet. Within this vast panorama of success I am especially interested in the devices that enabled them. Which particular features and stratagems led to such a triumph? Are they true human novelties, evolved from scratch to resolve human problems in hours of need, or are they actually retrofittings, part of solutions already available in the biological heritage?

In a search for such enabling devices, it is hardly

surprising that we begin by considering the human conscious mind itself. It looms large as an instrument potentially responsible for the traversal that brought our universe to its current eminence. That powerful human conscious mind has been assisted by remarkable learning and memory capacities and by extraordinary abilities to reason, decide, and create, all of which are complemented by language faculties in the verbal, mathematical, and musical domains. Thus richly equipped, humans would have been able to transition in record time from "plain beings" to "feeling and knowing beings." Little wonder then that they would have invented moral systems and religion, art, science and technology, politics and economics, and philosophy too, in brief, invented from scratch what we call, with our never satiated pride and presumption, human cultures. Having reshaped the earth to fit our goals—the biomass and the plain physical structure—humans would be about to do likewise to the contents of intergalactic space.

This account of how the conscious mind and the invention of human cultures would have helped us cope with the drama of life contains some obvious truths but also neglects important facts. Unfortunately, the omissions lead to a deformed interpretation of human achievements and predicaments and to a problematic account of the possible future.

The exaggerated distinction between human and nonhuman coping abilities, generated by an exceptionalist approach to human faculties, is deeply flawed. It is grandiose when it comes to humans; it unjustifiably diminishes nonhumans; and it fails to acknowledge the interdependence and cooperativity of living creatures, from the microscopic level to the human. Ultimately, it fails to acknowledge the presence of powerful *motifs, designs,* and *mechanisms* manifest in nature since life began—and even in the physics and chemistry that preceded it—and, in all likelihood, at least partly responsible for the blueprint of cultural developments usually attributed to humans.

A foundational motif is life itself, equipped with the set of chemical relationships and balances that permits *homeostasis* and the set of *homeostatic dictates* that helps identify perilous deviations from life-conducive ranges and commands the requisite corrections. All organisms, from bacteria to humans, rely on this foundational motif.

The designs and mechanisms that help support homeostatic demands are next on the list of humbling surprises. I am referring to intelligence, the ability to apply satisfactory solutions to problems posed by life, ranging from the procurement of basic sources of energy, such as nutrients and oxygen, to the control of territory and to defense against predation, along with strategies that address those problems, such as social cooperation and confrontation.

Once again the first and powerful example of such intelligences is present in bacteria. They solve with great ease all the problems in the above list. Their intelligence is non-explicit. It does not rely on minds with images of organism structure or images of the world around. It does not rely on feelings—barometers of the internal state of organisms—or on the consequent ownership of the organism and unique perspective that results from such ownership, in brief, on the phenomenon that we call consciousness. Yet the non-minded, non-explicit, hidden competence of such simple organisms has allowed their lives to be carried successfully over billions of years and offered a powerful design for the minded, explicit, overt intelligence that was to emerge in multicellular, brained creatures such as ourselves. The simple but far-reaching sensing/detecting ability featured in bacteria—or, for that matter, in plants—was the innovative mechanism that allowed simple organisms to detect stimuli such as temperature and the presence of others and react protectively and prospectively. Curiously, this modest debut of cognition was an anticipation of what overt feelings would later contribute in the setting of minds.

Minds, based on the mapping of overt, multi-dimensional patterns, were a powerful advance that permitted, simultaneously, making images of the world outside the organisms and images of the world inside them. The images of the exterior guided the successful actions of organisms in their environments, but feelings, the hybrid, interactive processes of the interior, at once mental and physical, were the most extraordinary enablers of adaptive and creative actions ever since nervous systems appeared on the scene, a mere 500 million years ago. They provided guidance and incentive to the creatures so equipped and founded consciousness too.

The appearance and structure of social phenomena and of the remarkable instruments of human cultures must be understood in the perspective of the biological phenomena that preceded them and enabled them. The long list of the latter includes *homeostatic regulation, non-explicit intelligences, sensing,* the *machinery for making images, feelings* as *mental translators of the life state within a com-*

plex organism, consciousness itself, and *mechanisms of social cooperation.* A powerful predecessor of the latter in the history of life is the "quorum sensing" ability of bacteria. As for a vivid example of the extraordinary consequences of interspecies cooperation, consider the human microbiome, where we find trillions of cooperating bacteria helping each of our individual human lives remain in good health while receiving from our own human life the support needed for their life cycle. Or, for that matter, consider the extraordinary cooperation to be found in forests, involving trees and fungi, below and above the ground.

By all means, we should indeed admire and even exalt the unique achievements of the human conscious mind and all the amazing novelty it created, over and above the solutions nature had already shepherded along. But we need to balance the account of how humans got to the present and recognize the fact that the fundamental devices we have used to succeed in our niche consist of transformations and upgrades of devices previously used by other living forms throughout a long history of individual and social successes. We need to respect the phenomenal and incompletely understood intelligence and designs of nature itself.

Behind the harmony or horror that we recognize in great art created by human intelligence and sensibility, there are related feelings of well-being, pleasure, suffering, and pain. Behind such feelings there are states of life that follow or violate the requirements of homeostasis. And underneath such life states there are chemical and physical process arrangements responsible for making life viable and for tuning the music of the stars and planets.

Acknowledging priorities and recognizing interdependence may come in handy as we cope with the ravages that we humans have inflicted on the earth and on its life, ravages that are likely responsible for some of the catastrophes we currently face, climate changes and pandemics being two prominent examples. It will give us an additional incentive to listen to the voices of those who dedicate their lives to thinking through the large-scale problems we face and recommend solutions that are wise, ethical, practical, and compatible with the big biological stage that humans occupy. There is some hope after all, and perhaps there should be some optimism as well.[1]

Notes and References

I: ON BEING

On Being, Feeling, and Knowing

1. In my previous book *The Strange Order of Things: Life, Feeling, and the Making of Cultures* (New York: Pantheon Books, 2018), I address the surprising facts discussed here. The very first creatures in the history of life were far more intelligent than one might have expected. See also Antonio Damasio and Hanna Damasio, "How Life Regulation and Feelings Motivate the Cultural Mind: A Neurobiological Account," in *The Cambridge Handbook of Cognitive Development*, ed. Olivier Houdé and Grégoire Borst (Cambridge, U.K.: Cambridge University Press, 2021), for a recent account of the intersection between biology and culture.

2. Quorum sensing is a striking example of the extraordinary intelligence of bacteria and of other single-cell organisms. See Stephen P. Diggle, Ashleigh S. Griffin, Genevieve S. Campbell, and Stuart A. West, "Cooperation and Conflict in Quorum-Sensing Bacterial Populations," *Nature* 450, no. 7168 (2007): 411–14; and Kenneth H. Nealson and J. Woodland Hastings, "Quo-

rum Sensing on a Global Scale: Massive Numbers of Bioluminescent Bacteria Make Milky Seas," *Applied and Environmental Microbiology* 72, no. 4 (2006): 2295–97.

The following provide details on life processes and on the extraordinary capabilities of single-cell organisms: Arto Annila and Erkki Annila, "Why Did Life Emerge?," *International Journal of Astrobiology* 7, no. 3–4 (2008): 293–300; Thomas R. Cech, "The RNA Worlds in Context," *Cold Spring Harbor Perspectives in Biology* 4, no. 7 (2012): a006742; Richard Dawkins, *The Selfish Gene: 30th Anniversary Edition* (New York: Oxford University Press, 2006); Christian de Duve, *Singularities: Landmarks in the Pathways of Life* (Cambridge, U.K.: Cambridge University Press, 2005); Christian de Duve, *Vital Dust: The Origin and Evolution of Life on Earth* (New York: Basic Books, 1995); Freeman Dyson, *Origins of Life* (New York: Cambridge University Press, 1999); Gerald Edelman, *Neural Darwinism: The Theory of Neuronal Group Selection* (New York: Basic Books, 1987); Gregory D. Edgecombe and David A. Legg, "Origins and Early Evolution of Arthropods," *Palaeontology* 57, no. 3 (2014): 457–68; Ivan Erill, Susana Campoy, and Jordi Barbé, "Aeons of Distress: An Evolutionary Perspective on the Bacterial SOS Response," *FEMS Microbiology Reviews* 31, no. 6 (2007): 637–56; Robert A. Foley, Lawrence Martin, Marta Mirazón Lahr, and Chris Stringer, "Major Transitions in Human Evolution," *Philosophi-*

cal Transactions of the Royal Society B 371, no. 1698 (2016), doi.org/10.1098/rstb.2015.0229; Tibor Gantí, *The Principles of Life* (New York: Oxford University Press, 2003); Daniel G. Gibson, John I. Glass, Carole Lartigue, Vladimir N. Noskov, Ray-Yuan Chuang, Mikkel A. Algire, Gwynedd A. Benders, et al., "Creation of a Bacterial Cell Controlled by a Chemically Synthesized Genome," *Science* 329, no. 5987 (2010): 52–56; Paul G. Higgs and Niles Lehman, "The RNA World: Molecular Cooperation at the Origins of Life," *Nature Reviews Genetics* 16, no. 1 (2015): 7–17; Alexandre Jousset, Nico Eisenhauer, Eva Materne, and Stefan Scheu, "Evolutionary History Predicts the Stability of Cooperation in Microbial Communities," *Nature Communications* 4 (2013); Gerald F. Joyce, "Bit by Bit: The Darwinian Basis of Life," *PLoS Biology* 10, no. 5 (2012): e1001323; Stuart Kauffman, "What Is Life?," *Israel Journal of Chemistry* 55, no. 8 (2015): 875–79; Daniel B. Kearns, "A Field Guide to Bacterial Swarming Motility," *Nature Reviews Microbiology* 8, no. 9 (2010): 634–44; Maya E. Kotas and Ruslan Medzhitov, "Homeostasis, Inflammation, and Disease Susceptibility," *Cell* 160, no. 5 (2015): 816–27; Karin E. Kram and Steven E. Finkel, "Rich Medium Composition Affects *Escherichia coli* Survival, Glycation, and Mutation Frequency During Long-Term Batch Culture," *Applied and Environmental Microbiology* 81, no. 13 (2015): 4442–50; Richard Leakey, *The Origin of Humankind* (New York: Basic Books, 1994); Derek

Le Roith, Joseph Shiloach, Jesse Roth, and Maxine
A. Lesniak, "Evolutionary Origins of Vertebrate Hormones: Substances Similar to Mammalian Insulins Are
Native to Unicellular Eukaryotes," *Proceedings of the
National Academy of Sciences* 77, no. 10 (1980): 6184–
88; Michael Levin, "The Computational Boundary of a
'Self': Developmental Bioelectricity Drives Multicellu-
larity and Scale-Free Cognition," *Frontiers in Psychology* (2019); Richard C. Lewontin, *Biology as Ideology:
The Doctrine of DNA* (New York: HarperPerennial,
1991); Mark Lyte and John F. Cryan, *Microbial Endo-
crinology: The Microbiota-Gut-Brain Axis in Health
and Disease* (New York: Springer, 2014); Alberto P.
Macho and Cyril Zipfel, "Plant PRRs and the Activation of Innate Immune Signaling," *Molecular Cell*
54, no. 2 (2014): 263–72; Lynn Margulis, *Symbiotic
Planet: A New View of Evolution* (New York: Basic
Books, 1998); Humberto R. Maturana and Francisco
J. Varela, "Autopoiesis: The Organization of Living," in
Autopoiesis and Cognition, ed. Humberto R. Maturana
and Francisco J. Varela (Dordrecht: Reidel, 1980), 73–
155; Margaret J. McFall-Ngai, "The Importance of
Microbes in Animal Development: Lessons from the
Squid-Vibrio Symbiosis," *Annual Review of Microbiology* 68 (2014): 177–94; Stephen B. McMahon, Federica
La Russa, and David L. H. Bennett, "Crosstalk Between
the Nociceptive and Immune Systems in Host Defense
and Disease," *Nature Reviews Neuroscience* 16, no. 7
(2015): 389–402; Lucas John Mix, "Defending Defini-

tions of Life," *Astrobiology* 15, no. 1 (2015): 15–19; Robert Pascal, Addy Pross, and John D. Sutherland, "Towards an Evolutionary Theory of the Origin of Life Based on Kinetics and Thermodynamics," *Open Biology* 3, no. 11 (2013): 130156; Alexandre Persat, Carey D. Nadell, Minyoung Kevin Kim, Francois Ingremeau, Albert Siryaporn, Knut Drescher, Ned S. Wingreen, Bonnie L. Bassler, Zemer Gitai, and Howard A. Stone, "The Mechanical World of Bacteria," *Cell* 161, no. 5 (2015): 988–97; Abe Pressman, Celia Blanco, and Irene A. Chen, "The RNA World as a Model System to Study the Origin of Life," *Current Biology* 25, no. 19 (2015): R953—R963; Paul B. Rainey and Katrina Rainey, "Evolution of Cooperation and Conflict in Experimental Bacterial Populations," *Nature* 425, no. 6953 (2003): 72–74; Kepa Ruiz-Mirazo, Carlos Briones, and Andrés de la Escosura, "Prebiotic Systems Chemistry: New Perspectives for the Origins of Life," Chemical Reviews 114, no. 1 (2014): 285–366; Erwin Schrödinger, *What Is Life?* (Cambridge, U.K.: Cambridge University Press, 1944); Vanessa Sperandio, Alfredo G. Torres, Bruce Jarvis, James P. Nataro, and James B. Kaper, "Bacteria-Host Communication: The Language of Hormones," *Proceedings of the National Academy of Sciences* 100, no. 15 (2003): 8951–56; Jan Spitzer, Gary J. Pielak, and Bert Poolman, "Emergence of Life: Physical Chemistry Changes the Paradigm," *Biology Direct* 10, no. 33 (2015); Eörs Szathmáry and John Maynard Smith, "The Major Evolutionary Transi-

tions," *Nature* 374, no. 6519 (1995): 227–32; D'Arcy
Thompson, *On Growth and Form* (Cambridge, U.K.:
Cambridge University Press, 1942); John S. Torday, "A
Central Theory of Biology," *Medical Hypotheses* 85,
no. 1 (2015): 49–57.

3. In a previous book, I addressed the notion of self, its
varieties, and considered their possible physiological
basis. Antonio Damasio, *Self Comes to Mind: Constructing the Conscious Brain* (New York: Pantheon,
2010).

II: ABOUT MINDS AND THE NEW ART
OF REPRESENTATION

Intelligence, Minds, and Consciousness

1. The work of František Baluška and Michael Levin is
especially relevant to the discussion of implicit intelligences. František Baluška and Michael Levin, "On
Having No Head: Cognition Throughout Biological
Systems," *Frontiers in Psychology* 7 (2016): 1–19;
František Baluška and Stefano Mancuso, "Deep Evo-
lutionary Origins of Neurobiology: Turning the Essence
of 'Neural' Upside-Down," *Communicative and Inte-
grative Biology* 2, no. 1 (2009): 60–65; František
Baluška and Arthur Reber, "Sentience and Conscious-
ness in Single Cells: How the First Minds Emerged in
Unicellular Species," *BioEssays* 41, no. 3 (2019); Paco
Calvo and František Baluška, "Conditions for Minimal
Intelligence Across Eukaryota: A Cognitive Science Per-

spective," *Frontiers in Psychology* 6 (2015): 1–4, doi
.org/10.3389/fpsyg.2015.01329.

Sensing Is Not the Same as Being Conscious and Does Not Require a Mind

1. Claude Bernard, *Leçons sur les phénomènes de la vie communs aux animaux et aux végétaux* (Paris: J.-B. Baillière et Fils, 1879), reprints from the collection of the University of Michigan Library; A. J. Trewavas, "What Is Plant Behaviour?," *Plant Cell and Environment* 32 (2009): 606–16; Edward O. Wilson, *The Social Conquest of the Earth* (New York: Liveright, 2012).

2. Colin Klein and Andrew B. Barron, "How Experimental Neuroscientists Can Fix the Hard Problem of Consciousness," *Neuroscience of Consciousness* 2020, no. 1 (2020): niaa009, doi.org/10.1093/nc/niaa009.

The Making of Mental Imagery

1. For a comprehensive review of their pioneering work on vision see David Hubel and Torsten Wiesel, *Brain and Visual Perception* (New York: Oxford University Press, 2004); Richard Masland, *We Know It When We See It: What the Neurobiology of Vision Tells Us About How We Think* (New York: Basic Books, 2020), provides a recent perspective on visual perception. See also Eric Kandel, James H. Schwartz, Thomas M. Jessell, Steven A. Siegelbaum, and A. J. Hudspeth, eds., *Principles of Neural Science,* 5th ed. (New York: McGraw-Hill, 2013); Stephen M. Kosslyn, *Image and Mind*

(Cambridge, Mass.: Harvard University Press, 1980); Stephen M. Kosslyn, Giorgio Ganis, and William L. Thompson, "Neural Foundations of Imagery," *Nature Reviews Neuroscience* 2 (2001): 635–42; Stephen M. Kosslyn, Alvaro Pascual-Leone, Olivier Felician, Susana Camposano, et al., "The Role of Area 17 in Visual Imagery: Convergent Evidence from PET and rTMS," *Science* 284 (1999): 167–70; Scott D. Slotnick, William L. Thompson, and Stephen M. Kosslyn, "Visual Mental Imagery Induces Retinotopically Organized Activation of Early Visual Areas," *Cerebral Cortex* 15 (2005): 1570–83.

2. The complexities of olfactory and gustatory perception have been investigated in the pioneering research of Richard Axel, Linda Buck, and Cornelia Bargmann. See, for example, L. Buck and R. Axel, "A Novel multigene family may encode odorant receptors: A molecular basis for odor recognition," *Cell* 65 (1991): 175-187.

Turning Neural Activity into Movement and Mind

1. Kandel, Schwartz, Jessell, Siegelbaum, and Hudspeth, *Principles of Neural Science.* Chapters concerning the anatomy and physiology of the nervous system.

Fabricating Minds

1. Stuart Hameroff, "The Quantum Origin of Life: How the Brain Evolved to Feel Good," in *On Human Nature,* ed. Michel Tibayrenc and Francisco José Ayala

(Amsterdam: Elsevier/AP, 2017), 333–53; Roger Penrose, "The Emperor's New Mind," *Royal Society for the Encouragement of Arts, Manufactures, and Commerce* 139, no. 5420 (1991): 506–14, www.jstor.org/stable/41378098.

The Minds of Plants and
the Wisdom of Prince Charles

1. Walter B. Cannon, *The Wisdom of the Body* (New York: Norton, 1932); Walter B. Cannon, "Organization for Physiological Homeostasis," *Physiological Review* 9 (1929): 399–431; Claude Bernard, *Leçons sur les phénomènes de la vie communs aux animaux et aux végétaux* (Paris: J.-B. Baillière et Fils, 1879), reprints from the collection of the University of Michigan Library; Michael Pollan, "The Intelligent Plant," *New Yorker,* Dec. 23 and 30, 2013.

2. In certain circumstances, plants can be part of collaborative and even symbiotic relationships. The underground networks of tree roots in forests are prime examples. All demonstrate the power of unminded, non-conscious, and, needless to say, non-neural varieties of intelligence. See Monica Gagliano, *Thus Spoke the Plant* (New York: Penguin Random House, 2018).

Algorithms in the Kitchen

1. Michel Serres, *Petite Poucette* (Paris: Le Pommier, 2012).

III: ON FEELINGS

The Beginnings of Feeling: Setting the Stage

1. Stuart Hameroff, among others, has suggested that organisms might have had feelings before nervous systems ever appeared. The source for this idea, as I understand it, is the fact that certain "physical configurations" are more likely to be associated with more stable and viable life states. I believe that is indeed the case, but it does not follow that such conducive physical configurations would or could generate feelings, that is, generate mental states concerning the current condition of the organism. To the best of my understanding, the existence of mental states requires the presence of considerable elaborate nervous systems and depends on the representation of organism states in neural maps. See Stuart Hameroff, "The Quantum Origin of Life: How the Brain Evolved to Feel Good," in *On Human Nature*, ed. Michel Tibayrenc and Francisco José Ayala (Amsterdam: Elsevier/AP, 2017), 333–53.

Affect

1. My use of the term "primordial" is conventional and meant to refer to the simple and direct nature of what I conceive of feelings as having been as they emerged in early human evolution and as they still are likely to be in many nonhuman species not to mention human infants. I refer to all such early feelings as "homeo-

static" to separate them clearly from emotional feelings whose source is the engagement of emotions. Derek Denton has authored an important book titled *The Primordial Emotions,* where the term "primordial" points to the class of homeostatic processes that produce, in his words, "imperious states of arousal and compelling intentions to act." Breathing and excretion processes (for example, micturition) provide the setting. These primordial emotions are followed by the respective feelings. The prime situation causing such primordial emotions/feelings is the blockage of respiratory passages and the resulting "air-hunger." Derek Denton, *The Primordial Emotions: The Dawning of Consciousness* (Oxford: Oxford University Press, 2005).

2. Manos Tsakiris and Helena De Preester have assembled a remarkable collection of articles on the topic of interoception authored by most of the neuroscience leaders currently interested in interoception: *The Interoceptive Mind: From Homeostasis to Awareness,* ed. Manos Tsakiris and Helena De Preester (Oxford: Oxford University Press, 2019).

See also A. D. Craig, *How Do You Feel? An Interoceptive Moment with Your Neurobiological Self* (Princeton, N.J.: Princeton University Press, 2015); A. D. Craig, "Interoception: The Sense of the Physiological Condition of the Body," *Current Opinion in Neurobiology* 13, no. 4 (2003): 500–505; Hugo D. Critchley, Stefan Wiens, Pia Rotshtein, Arne Öhman, and Raymond J. Dolan, "Neural Systems Supporting Interoceptive Awareness," *Nature Neuroscience* 7, no. 2 (2004): 189–95.

3. For a reasonable distinction between homeostasis and allostasis, see Bruce S. McEwen, "Stress, Adaptation, and Disease: Allostasis and Allostatic Load," *Annals of the New York Academy of Sciences* 840, no. 1 (1998): 33–44.

4. The following sources cover the topic of affect quite extensively, ranging from general conception to neural biological implementation: Ralph Adolphs and David J. Anderson, *The Neuroscience of Emotion: A New Synthesis* (Princeton, N.J.: Princeton University Press, 2018); Ralph Adolphs, Hanna Damasio, Daniel Tranel, Greg Cooper, and Antonio Damasio, "A Role for Somatosensory Cortices in the Visual Recognition of Emotion as Revealed by Three-Dimensional Lesion Mapping," *Journal of Neuroscience* 20, no. 7 (2000): 2683–90; Antonio Damasio, *The Feeling of What Happens: Body and Emotion in the Making of Consciousness* (New York: Harcourt Brace, 1999); Antonio Damasio, Hanna Damasio, and Daniel Tranel, "Persistence of Feelings and Sentience After Bilateral Damage of the Insula," *Cerebral Cortex* 23 (2012): 833–46; Antonio Damasio, Thomas J. Grabowski, Antoine Bechara, Hanna Damasio, Laura L. B. Ponto, Josef Parvizi, and Richard Hichwa, "Subcortical and Cortical Brain Activity During the Feeling of Self-Generated Emotions," *Nature Neuroscience* 3, no. 10 (2000): 1049–56,

doi.org/10.1038/79871; Antonio Damasio and Joseph LeDoux, "Emotion," in *Principles of Neural Science,* ed. Eric Kandel, James H. Schwartz, Thomas M. Jessell, Steven A. Siegelbaum, and A. J. Hudspeth, 5th ed. (New York: McGraw-Hill, 2013); Richard Davidson and Brianna S. Shuyler, "Neuroscience of Happiness," in *World Happiness Report 2015,* ed. John F. Helliwell, Richard Layard, and Jeffrey Sachs (New York: Sustainable Development Solutions Network, 2015); Mary Helen Immordino-Yang, *Emotions, Learning, and the Brain: Exploring the Educational Implications of Affective Neuroscience* (New York: W. W. Norton, 2015); Kenneth H. Nealson and J. Woodland Hastings, "Quorum Sensing on a Global Scale: Massive Numbers of Bioluminescent Bacteria Make Milky Seas," *Applied and Environmental Microbiology* 72, no. 4 (2006): 2295–97; Anil K. Seth, "Interoceptive Inference, Emotion, and the Embodied Self," *Trends in Cognitive Sciences* 17, no. 11 (2013): 565–73; Mark Solms, *The Feeling Brain: Selected Papers on Neuropsychoanalysis* (London: Karnac Books, 2015); Anthony G. Vaccaro, Jonas T. Kaplan, and Antonio Damasio, "Bittersweet: The Neuroscience of Ambivalent Affect," *Perspectives on Psychological Science* 15 (2020): 1187–99.

Biological Efficiency
and the Origin of Feelings

1. Stuart Hameroff, "The Quantum Origin of Life: How the Brain Evolved to Feel Good," in *On Human*

Nature, ed. Michel Tibayrenc and Francisco José Ayala (Amsterdam: Elsevier/AP, 2017), 333–53.

Grounding Feelings III

1. Helena De Preester has written an incisive and informative piece on the phenomenology of feeling that concerns this issue directly. Feelings, if we must refer to them as "perceptions," are certainly *unconventional* examples of such processes. Helena De Preester, "Subjectivity as a Sentient Perspective and the Role of Interoception," in Tsakiris and De Preester, *Interoceptive Mind.*

Grounding Feelings IV

1. Antonio Damasio and Gil B. Carvalho, "The Nature of Feelings: Evolutionary and Neurobiological Origins," *Nature Reviews Neuroscience* 14, no. 2 (2013): 143–52; Gil Carvalho and Antonio Damasio, "Interoception as the Origin of Feelings: A New Synthesis," *BioEssays* (forthcoming June 2021).

Grounding Feelings V

1. Antonio Damasio, *The Strange Order of Things: Life, Feeling, and the Making of Cultures* (New York: Pantheon Books, 2018).

Grounding Feelings VI

1. Derek Denton, *Primordial Emotions: The dawning of consciousness* (Oxford: Oxford University Press, 2005).

Grounding Feelings VII

1. He-Bin Tang, Yu-Sang Li, Koji Arihiro, and Yoshihiro Nakata, "Activation of the Neurokinin-1 Receptor by Substance P Triggers the Release of Substance P from Cultured Adult Rat Dorsal Root Ganglion Neurons," *Molecular Pain* 3, no. 1 (2007): 42, doi.org/10.1186/1744-8069-3-42.

Homeostatic Feelings in a Sociocultural Setting

1. The deep connections between biological phenomena and sociocultural structures and operations is discussed in *Strange Order of Things* (cited before). See also Marco Verweij and Antonio Damasio, "The Somatic Marker Hypothesis and Political Life" in *Oxford Research Encyclopedia of Politics* (Oxford University Press, 2019).

IV: ON CONSCIOUSNESS AND KNOWING

Why Consciousness? Why Now?

1. I provide an account of the close relation between biology and the evolution of cultures in my book *The Strange Order of Things: Life, Feeling, and the Making of Cultures* (New York: Pantheon Books, 2018).

2. W. H. Auden, *For the Time Being: A Christmas Oratorio* (London: Plough, 1942).

Natural Consciousness

1. The word "consciousness" is so recent that it does not figure in Shakespeare at all. Romance languages never developed an equivalent for the English word "consciousness" and still use "conscious" both as a synonym of "consciousness" and in reference to moral behavior. When Hamlet says, "Thus conscience does make cowards of us all," he is referring to moral qualms and not to consciousness. The word "consciousness" makes its appearance in 1690, defined by John Locke as "the perception of what passes in a man's mind." Not too bad but not as good as it needs to be.

2. Derek Denton. *The Primordial Emotions: The Dawning of Consciousness* (Oxford: Oxford University Press, 2005).

The Problem of Consciousness

1. Stuart Hameroff and Christof Koch are two biologists who have adopted a panpsychic perspective in their work on consciousness.

2. David J. Chalmers, *The Conscious Mind: In Search of a Fundamental Theory* (Oxford: Oxford University Press, 1996).

3. Thomas Nagel, "What Is It Like to Be a Bat?," *Philosophical Review* 83, no. 4 (1974): 435–50, doi.org/10.2307/2183914.

4. A number of philosophers have criticized the hard problem position for other reasons, as is the case with Daniel Dennett. Daniel C. Dennett, "Facing Up to the

Hard Question of Consciousness," *Philosophical Transactions of the Royal Society B* (2018), doi.org/10.1098/rstb.2017.0342.

5. For a recent review of theories and facts concerning consciousness, see Simona Ginsburg and Eva Jablonka, *The Evolution of the Sensitive Soul: Learning and the Origins of Consciousness* (Cambridge, Mass.: MIT Press, 2019). It offers a comprehensive survey of contemporary views on consciousness encompassing both primarily physiological and biological perspectives. See also Antonio Damasio "Feeling & Knowing: Making Minds Conscious," *Cognitive Neuroscience* (2021).

Being Conscious Is Not the Same as Being Awake

1. Antonio Damasio and Kaspar Meyer, "Consciousness: An Overview of the Phenomenon and of Its Possible Neural Basis," in *The Neurology of Consciousness,* ed. Steven Laureys and Giulio Tononi (Burlington, Mass.: Elsevier, 2009), 3–14.

Extended Consciousness

1. Antonio Damasio, *The Feeling of What Happens: Body and Emotion in the Making of Consciousness* (New York: Harcourt Brace, 1999).

With Ease, and You Beside

1. Emily Dickinson, "Poem XLIII," in *Collected Poems* (Philadelphia: Courage Books, 1991).

A Gathering of Knowledge

1. My colleague Max Henning commented on the above passage as follows: "Accounting for consciousness by locating the mental subject not in some special and distinct physiological function or substance, but rather piecemeal in attributes of every image in the mental flow, has an intriguing precedent in Buddhist philosophy. Specifically, the Buddhist doctrines of 'non-self' (*anattā* in Pali) and 'dependent origination' hold that the mental subject or 'self' has no distinct substantive essence; it exists only in relation to mental 'objects,' which in turn exist only in relation to the subject as the philosopher David Loy suggests. This apparent convergence of soteriological and epistemological inquiry on the nature of consciousness and the mental subject invites further investigation."

   David R. Loy, *Nonduality: In Buddhism and Other Spiritual Traditions* (Wisdom Publications, 2019).

Integration Is Not the Source of Consciousness

1. Giulio Tononi and Christof Koch have advanced a different role for the integration of information. See Christof Koch, *The Feeling of Life Itself: Why Consciousness Is Widespread but Can't Be Computed* (Cambridge, Mass.: MIT Press, 2019). The word "feeling" in Koch's book title apparently refers to a conjunction of cognitive factors and not to the affective phenomenon I discuss in this book.

Consciousness and Attention

1. Stanislas Dehaene and Jean-Pierre Changeux have contributed remarkably to elucidating the intersection of attention and consciousness and provided the fundamental texts in this area. See Stanislas Dehaene, *Consciousness and the Brain: Deciphering How the Brain Codes Our Thoughts* (New York: Viking, 2014).

Loss of Consciousness

1. Personal recollections.

2. František Baluška, Ken Yokawa, Stefano Mancuso, and Keith Baverstock, "Understanding of Anesthesia—Why Consciousness Is Essential for Life and Not Based on Genes," *Communicative and Integrative Biology* 9, no. 6 (2016), doi.org/10.1080/19420889.2016.1238118.

3. Jerome B. Posner, Clifford B. Saper, Nicholas D. Schiff, and Fred Plum, *Plum and Posner's Diagnosis of Stupor and Coma* (New York: Oxford University Press, 2007).

4. See Damasio, *Feeling of What Happens,* chapter 8 on the neurology of consciousness. See also Josef Parvizi and Antonio Damasio, "Neuroanatomical Correlates of Brainstem Coma," *Brain* 126, no. 7 (2003): 1524–36; Josef Parvizi and Antonio Damasio, "Consciousness and the Brainstem," *Cognition* 79, no. 1 (2001): 135–60.

The Cerebral Cortices and the Brain Stem in the Making of Consciousness

1. Antonio Damasio, *Self Comes to Mind: Constructing the Conscious Brain* (New York: Pantheon, 2010);

Antonio Damasio, Hanna Damasio, and Daniel Tranel, "Persistence of Feelings and Sentience After Bilateral Damage of the Insula," *Cerebral Cortex* 23 (2012): 833–46; Antonio Damasio and Kaspar Meyer, "Consciousness: An Overview of the Phenomenon and of Its Possible Neural Basis," in *The Neurology of Consciousness,* ed. Steven Laureys and Giulio Tononi (Burlington, Mass.: Elsevier, 2009), 3–14.

Feeling Machines and Conscious Machines

1. Kingson Man and Antonio Damasio, "Homeostasis and Soft Robotics in the Design of Feeling Machines," *Nature Machine Intelligence* 1 (2019): 446–52, doi.org/10.1038/s42256-019-0103-7.

V: IN ALL FAIRNESS: AN EPILOGUE

1. The ideas of Peter Singer and Paul Farmer are examples of responses to humanity's current predicaments that I especially admire. See Peter Singer, *The Expanding Circle: Ethics, Evolution, and Moral Progress* (Princeton, N.J.: Princeton University Press, 2011); Paul Farmer, *Fevers, Feuds, and Diamonds: Ebola and the Ravages of History* (New York: Farrar, Straus and Giroux, 2020).

Other Readings

Lawrence W. Barsalou. "Grounded Cognition." *Annual Review of Psychology* 59 (2008): 617–45.

Nick Bostrom. *Superintelligence: Paths, Dangers, Strategies.* Oxford: Oxford University Press, 2014.

Sean Carroll. *The Big Picture.* New York: Dutton, 2016.

John Gray. *The Silence of Animals: On Progress and Other Modern Myths.* New York: Farrar, Straus and Giroux, 2013.

Siri Hustvedt. *The Delusions of Certainty.* New York: Simon & Schuster, 2017.

Rodrigo Quian Quiroga. "Plugging into Human Memory: Advantages, Challenges, and Insights from Human Single-Neuron Recordings." *Cell* 179, no. 5 (2019): 1015–32. doi.org/10.1016/j.cell.2019.10.016.

David Rudrauf, Daniel Bennequin, Isabela Granic, Gregory Landini, Karl Friston, and Kenneth Williford. "A Mathematical Model of Embodied Consciousness." *Journal of Theoretical Biology* 428 (2017): 106–31. doi.org/10.1016/j.jtbi.2017.05.032.

John S. Torday. "A Central Theory of Biology." *Medical Hypotheses* 85, no. 1 (2015): 49–57.

Luis P. Villarreal. "Are Viruses Alive?" *Scientific American* 291, no. 6 (2004): 100–105. doi.org/10.2307/26 060805.

Edward O. Wilson. *The Social Conquest of the Earth*. New York: Liveright, 2012.

Acknowledgments

This is the space where authors usually describe the circumstances under which their particular project was born. In the case of this book, however, I have already explained in the preface how an idea of my editor, Dan Frank, and my own frustration with the traditional science book format, guided me to *Feeling and Knowing*. My thanks to him for placing me on the path of rediscovering my own work and of realizing that I had actually solved some of the scientific problems over which I agonized.

This is also the space meant to recognize the colleagues and friends who make this kind of bizarre pursuit possible. I will first mention my colleagues at the Brain and Creativity Institute, with whom I live my scientific day to day exchanging ideas on all aspects of biology, psychology, and neuroscience. Some of them patiently read first versions of the manuscript, commented intelligently, and advised wisely. They are Kingson Man, Jonas Kaplan, Max Henning, Helder Araujo, Anthony Vacarro, John Monterosso, Marco

Verweij, Gil Carvalho, Assal Habibi, Rael Cahn, Mary Helen Immordino-Yang, Leonardo Christov-Moore, Morteza Dehghani, and Lisa Aziz Zadeh.

A number of friends were kind enough to read, to encourage and comment. They include Peter Sacks, Jorie Graham, Hartmut Neven, Nicolas Berggruen, Dan Tranel, Josef Parvizi, Barbara Guggenheim, Regina Weingarten, Julian Morris, Landon Ross, Silvia Gaspardo Moro, and Charles Ray. My gratitude, all the more so since this is not the first time that some of them kept me company in the foolish effort of putting down ideas on a page.

I have noticed, over the years, that the writing of my books depends on the stability of the work environment and that the music I hear and the art I see become associated with the work to the point of being necessary for its continuation. I know that some of my previous books are indelibly connected to Maria João Pires, Yo-Yo Ma, and Daniel Barenboim, among others, admired artists and friends. This time the cellist Elena Andreyev and her rich version of the Bach Cello Suites became an island of stability and clarity in many hours of need. I am grateful for her company.

Michael Carlisle and Alexis Hurley are not only the most professional literary agents but also indispensable friends. My thanks to them for their good humor and encouragement.

I can no longer imagine my professional life without Denise Nakamura. She is the calmest and most competent of office managers, she conducts bibliographic searches with a calm few of us ever can muster and she prepares both my handwritten and dictated texts flawlessly. No amount of thanks will suffice for her.

Hanna Damasio knows what I think but still reads every word that I write. Whether or not she agrees with my ideas, she comments patiently and constructively. Her contributions are central to the work, and my gratitude is immense.

Index

Page numbers followed by *f* and *t* refer to figures and tables, respectively.

A Note About the Author

Antonio Damasio is Dornsife Professor of Neuroscience, Psychology and Philosophy, and Director of the Brain and Creativity Institute at the University of Southern California in Los Angeles. Trained as both neurologist and neuroscientist, Damasio has made seminal contributions to the understanding of affect, decision-making, and consciousness, and had a major impact in neuroscience, psychology, and philosophy. He is one of the most highly cited scientists worldwide. His most recent work addresses the role of homeostasis in robotic design.

Damasio is a member of the National Academy of Medicine and a Fellow of the American Academy of Arts and Sciences. He has received numerous prizes, among them the Pessoa Prize, the Prince of Asturias Prize in Science and Technology, the Grawemeyer Award, the Honda Prize, the International Freud Medal, and the Paul MacLean Award. He holds Honorary Doctorates from several leading universities, some shared with his wife, Hanna, among them the École Polytechnique Fédérale de Lausanne and the Sorbonne.

dornsife.usc.edu/bci

antoniodamasio.com

A Note on the Type

The text of this book was set in Sabon, a typeface designed by Jan Tschichold (1902–1974), the well-known German typographer. Designed in 1966 and based on the original designs by Claude Garamond (ca. 1480–1561), Sabon was named for the punch cutter Jacques Sabon, who brought Garamond's matrices to Frankfurt.

Typeset by Scribe, Philadelphia, Pennsylvania

Printed and bound by Berryville Graphics,
Berryville, Virgina